LEAN Production – Easy and Comprehensive

Roman Hänggi · André Fimpel ·
Roland Siegenthaler

LEAN Production – Easy and Comprehensive

A practical guide to lean processes
explained with pictures

Springer Vieweg

Roman Hänggi
Brülisau, Switzerland

André Fimpel
Stuttgart, Germany

Roland Siegenthaler
Meilen, Switzerland

ISBN 978-3-662-64526-0   ISBN 978-3-662-64527-7   (eBook)
https://doi.org/10.1007/978-3-662-64527-7

Responsible Editor: Michael Kottusch
This Springer imprint is published by the registered company Springer-Verlag GmbH, DE part of Springer Nature.
The registered company address is: Heidelberger Platz 3, 14197 Berlin, Germany

Foreword

The starting point of "Lean Production – simple and comprehensive" was following Roland's presentation on "Visualizing Complex Issues" when we talked about the complexity of implementing Lean Management. For years, we had been driven by the question of how to communicate Lean theory simply and broadly. After all, a comprehensive understanding of Lean is needed at the beginning of every successful Lean change. True to the motto "just do it", we started to combine our knowledge and skills on that and brought all of our practical experience together. We put the Lean principles into metaphors and illustrated them with pictures. At that moment, the idea for "Lean Production – Simple and Comprehensive" was born. A Lean professor, a Lean practice expert and an illustrator had come together to create an exciting path.

After almost three years with many controversial discussions, harmonious work meetings and many night shifts, we have now finalized the first edition of "Lean Production – Simple and Comprehensive". The search for the right metaphors, the logical structure, the most coherent stories and self-speaking images was not always easy and often challenged us. The feedback from many friends and colleagues has always motivated us. We appreciated their advice and their input moved the book forward.

Our goal is to always keep the topics *simple,* using pictures, examples and concrete experiences from many Lean projects to give the reader a vivid idea about Lean. On the other hand, we present Lean *comprehensively* and propose a concrete guide that has been proven in practice. We cover the entire spectrum from waste basics to the most important Lean principles to field-tested Lean methods and Lean change.

We hope to inspire both the Lean beginner and the Lean expert to find new ways to eliminate waste from their processes every day. And we hope you enjoy reading it.

Roman Hänggi   André Fimpel   Roland Siegenthaler

Introduction

What Is Lean Production?

Lean Production means producing without waste and without detours. The goal is to ensure quality, punctuality and productivity as well as to create the conditions for automation and digitalization.

In a Lean ideal process, the customer is the focus and always gets the right product, at the right place, at the right time, in the right quantity and in the right quality. This 5 X Right basic principle, also known as the 5R principle or just-in-time principle, is a central pillar in the Toyota Production System, the origin of Lean Production (Ōno et al. 2013).

Everything that stands in the way of implementing the just-in-time principle is waste in the sense of Lean Production. Several Lean principles and methods exist to eliminate this waste.

To go down this new path, not only is the knowledge of methods and tools are important, but a change in thinking and a new corporate culture are necessary. This change is the most difficult step. We want to inspire you with this book to follow the path of Lean transformation.

Isn't Lean Out?

The first thoughts about Lean Production were made in the middle of the last century in the Toyota factories. In the 1990s, Lean became known in our latitudes through several publications (Womack et al. 2007) and a wave of enthusiasm started, which, however, has noticeably decreased in recent years. Over the years, Lean was labeled as old-fashioned and slowly forgotten. From our point of view, Lean Production has unfortunately only been properly understood in a few companies and therefore only been implemented consistently in individual cases over the years.

Quite often, only a few methods from the Lean toolbox are applied selectively, without pursuing an overall goal and vision. A true Lean culture is missing.

But Lean is like buckling up in a car or washing your hands. It may no longer be new, but it has no expiration date. It is and remains important, meaningful and effective. Today more than ever.

Digitalization Brings Lean Back Into Play!

As we will show with some examples, when using the many available technical possibilities, there is a danger that technologies are used to control waste instead of eliminating it. In all the euphoria surrounding digitalization, Lean therefore takes on great strategic importance for every company today more than ever before. Only when processes are free of waste does digitalization take flight.

This motivated us to bring individual Lean fragments back into an overall context, but still package them clearly. The book presents the interrelationships around Lean in its entirety. This will help you to optimize your production comprehensively and to counter the critical voices in the Lean implementation. The success of your measures will be the engine for the Lean culture and lay the foundation for the digitalization (then making sense) in your company.

Lean Also Works Without a Car Factory

By the way: unfortunately, there is still a persistent belief that Lean Production only works in the automotive industry with large production lines and millions of units. The Toyota engineer Taiichi Ohno (Ōno and Bodek 2008) developed the basics in this context, and we have also gained our experience with Lean, especially in production operations, but not only in automotive manufacturing. However, Lean Production just means "to produce without waste". And this also works in the bakery, on the construction site, in the office or even at home in the kitchen.

Espresso Without Waste

Start by Seeing the Waste

If you examine a process for waste, for example the preparation of espresso, you will be astonished and shocked to discover that the barista spends 45 min per hour waiting and doing other useless things. Only 15 min, or 25% of his capacity, are invested in value creating activities that bring benefits to the customer. If you want to optimize this process, it is not about working faster, but about removing the waste from the process. Making the waste visible and quantifying, it is the first important step on the way to a Lean process.

So where is the waste hiding in the espresso process? To see this, you have to mentally break down the process into small sub-steps: the barista takes the cup (1), carries it to the coffee machine (2), presses start (3), waits until the cup is full (4), and finally the espresso is ready!

First that the barista has to carry the cup the distance from the cupboard to the machine. In the true meaning of Lean, this is waste. Waiting during the coffee making process is also not value creating and therefore waste. So even in this small example, there is quite a bit of waste hidden.

Shortcut Lean through Digitalization?

No matter what kind of waste is currently plaguing us in the production, digitalization seems to be the elegant solution for everything today! But even if you buy the most modern connected robot that autonomously carries your cup from the cupboard to the coffee machine, the waste is still there. Now automated.

The Way to Lean Produced Espresso

"Don't digitize your mess!" is therefore the top motto. Anyway, it is much cheaper to push the cupboard to the machine anyway than to buy and program a robot. And that is the only way to really eliminate waste.

This is what Lean is all about: seeing waste and systematically eliminating it from your process.

Understanding Lean in Five Steps

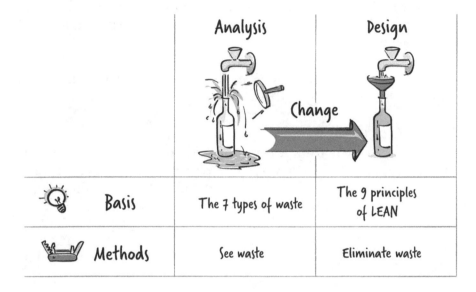

		Analysis	Design
		The 7 types of waste	The 9 principles of LEAN
	Basis		
	Methods	See waste	Eliminate waste

Lean theory is only simple at first glance. Our experience shows that there is a lot of confusion regarding fundamentals, terminology and content. Our structure of the book will remedy this.

After this introduction, the first topic is theory: understanding waste and how to avoid it.

Chap. 1 – The 7 Types of Waste:
Lean Production is waste-free production. So first you need to know what waste is and what types of it there are.

Chap. 2 – The 9 Principles to Avoid Waste:
To eliminate waste and its causes, you need to align your processes with Lean principles. We present the nine most important ones.

Now it is getting practical: Chaps. 3 and 4 deal with concrete methods and tools to see the waste and design your production according to the Lean principles.

Chap. 3 – Methods to See Waste:
Here we show you methods to see and quantify waste in your production.

Chap. 4 – Methods to Eliminate Waste:
Now that you can see waste in your processes, you need to eliminate it. We will show you the most important methods from the Lean toolbox that you can use to implement the 9 principles and thus sustainably banish the 7 types of waste from your processes.

Finally, we discuss approaches to implementing the change to a Lean company.

Chap. 5 – Lean Change:
To become a Lean company, it is unfortunately not enough to apply some methods selectively. Lean only works if it is properly anchored in the company. Lean change requires a comprehensive cultural change. This must be carefully considered and implemented. We show how it can succeed.

Contents

About the Authors

Roman Hänggi Professor of Production Management
After studying engineering and earning a doctorate in economics, Roman started in industry in the early 1990s, optimizing an optics manufacturing facility at Leica with Lean. This enthusiasm for Lean has accompanied him throughout his professional life. He implemented further successful Lean projects in production at Bosch, Hilti and Arbonia and used this experience from production to also optimize processes in development, service or administration with Lean. Curiosity and the desire to pass on his extensive knowledge from industry led him to the Chair of Production Management at the University of Applied Sciences OST. There he motivates students in Rapperswil and St. Gallen in lectures on Lean and digitalization in industry. Practice is important to him. That is why he supports industrial companies

on their way to becoming Lean champions and Industry 4.0 winners. He also teaches as a lecturer in executive programs at the University of St. Gallen. In his free time, Roman can be found on the ski slopes in Appenzell (Switzerland).

André Fimpel Lean Manager & Lean Consultant

Lean is a question of culture. In addition, when it comes to culture, André has a broad horizon. Born in Brazil and raised in Germany and Argentina, he knows all facets of people's ways of living and working. After graduating as an industrial engineer, he started his mission Lean as a consultant 15 years ago: first for the Fraunhofer Institute for Manufacturing Engineering and Automation in Stuttgart, then for Porsche Consulting and now for Hermes Schleifmittel GmbH. However, consultant is probably the wrong name for his vocation. He is a Lean enthusiast, a tinkerer and a communicator who manages to get both the mechanic at the Chinese car plant and the CEO of the Italian shoe factory excited about Lean Production. André is the most tolerant person on earth, except for one thing he hates with all his heart: inventories and the six other types of waste.

Roland Siegenthaler Illustrator for knowledge and processes
Roland is extremely curious, extremely creative and extremely comfortable. These three characteristics have qualified him for the post of questioner, draftsman and text seasoner in this book. Actually, Roland is an electrical engineer. At the age of 30, however, he realized that projects rarely fail because of technology, but mostly because of communication. As an autodidact, he has therefore developed into a visualization professional over the years. With his explanatory skills, he now supports teams, from production to management, in communicating innovation and change. His clientele ranges from small businesses to large international corporations. Roland's audience also includes his three funny children, to whom he tries to explain this fascinating and crazy world with pictures.

Abbreviations

5S	Lean method, which organizes the workplace with the five steps of sorting, cleaning, making visible, standardizing and (ab)securing the standard
5R	5X-Right basic principle or also just-in-time principle. Means ensuring the right part, in the right quality, at the right time, in the right place, in the right quantity
A3	Method of structuring the problem solution on an A3 sheet of paper
ABC–XYZ Analysis	Classification of parts according to part price for purchased parts or manufacturing costs for in-house production (ABC classification) and consumption (XYZ classification)
Andon	Method from the Toyota production system for immediate visualization of process problems, e.g., using a signal lamp
BOM	Bill of Material, the structure or tree of a product, list of material of assemblies, subassemblies and single parts of a product, often the term parts list is being used
Clock	Time in which a process is repeated
C-Parts	Low value materials (such as small parts, e.g., screws, nuts)
CT	Cycle time, time between completion of two products, time a part takes to go through a process
ERP	Enterprise resource planning, IT system for planning and controlling the flow of goods and values within a company.

FIFO	First-in-first-out, parts produced or stored first are removed first
Gemba	Japanese, "the real place". In the Lean context, this is the place where value creation happens
Handling Step	Processes such as transporting, testing, unpacking, storing or even transferring are handling steps and waste
Handling Step Analysis	The handling step analysis presents the handling steps for a process in context and helps to see the waste
Industry 4.0	Use of digital technologies to increase productivity in an industrial company, especially in production
Jidoka	Intelligent mechanical solutions to avoid errors, path to full automation
JIS	Just-in-sequence, external delivery in sequence
JIT	Just-in-time, direct delivery between processes without warehouse and intermediate buffer
Kaizen	Management approach to continuous improvement
Kanban	Methods for pull control, based on a signal(=Kanban) for replenishment control
KPI	Key performance indicator, measure of key performance parameter in the company
LS, lot size	Related production or procurement of parts
Lead Time	Time a part takes to go through one or multiple operation processes
Milkrun	Clocked route trains to supply production at specified intervals
MRP	Material requirement planning, term for planning and control via BOM explosion and inventory reconciliation.
MRPII	MRP including capacity planning
MTM	Methods–time–measurement, method for time evaluation based on time modules with standardized and defined specifications
Muda	Japanese term for waste
OBC	Operator balance chart, diagram for visualization of work distribution
OEE	Overall equipment effectiveness, method for measuring the effectiveness of equipment or processes.
Pareto rule	80–20 rule means 20% of causes/issues cause 80% of problems/results
PDCA	Approach from quality management for continuous problem solving in four phases ("plan"–"do"–"check"–"act")

Poka-Yoke	Method for avoiding errors through technical or organizational measures on the process or product.
PPS System	Production planning and control system, computer program or system that plans and controls production centrally according to push control.
Predictive Maintenance	Predictive maintenance, especially of production machines, relevant use case in learning from data in the factory of tomorrow
Pull	MRP type with defined stock that cannot be exceeded, always triggered by customer consumption
Push	Centrally planned disposition, independent of the maximum stock and the current demand of the next process
PT	Processing time, time the machine takes to manufacture one part, real time of production without any setup time or waiting time
REFA	Method of the association work design, business organization and enterprise development for the determination of time
Sequencing	Sequencing is the sequence planning of the orders
Shopfloor Management	In short daily meetings, the status of production is discussed and, if necessary, measures are defined.
SMED	Single-minute exchange of die, method for setup time reduction
Spaghetti Diagram	Graphical representation of the paths in workstations, the representation of the complex paths resembles "spaghettis".
ST	Setup time, time to prepare a machine or process for manufacturing.
Takt	Customer cycle (customer takt), process cycle derived from customer demand
Waste	Excessive consumption or inefficient use of resources in processes
Waste Walk	Structured walk-through of production to identify waste
WT	Available working time of the processes
Yield	Ratio of good parts to produced parts in %.
Zoning	Arrangement of materials, work equipment or storage locations according to frequency and priority

1

The 7 Types of Waste

Contents

Muda (無駄, on'yomi reading) is a Japanese word meaning "futility; use-lessness; wastefulness",[1] and is a key concept in Lean process thinking (WIKIPEDIA)

1.1 Why Do We Distinguish Waste?

We have seen from the simple example of our barista that the largest part of a process can be waste. In Lean Management, it has become established to categorize waste into different types. We talk about "types" of waste. This helps to find, quantify and eliminate it more specifically in the processes.

© Springer-Verlag GmbH Germany, part of Springer Nature 2022
R. Hänggi et al., *LEAN Production – Easy and Comprehensive*,
https://doi.org/10.1007/978-3-662-64527-7_1

For Lean Management understanding waste and its nature is like the 1 × 1 in mathematics. Everything is based on it. Therefore, in this chapter, we will take a closer look at the 7 types of waste. Once you understand them, with some practice you can develop an eye and a sense for them.

The Seven Types of Waste

1. Overproduction
2. Stock
3. Transport
4. Motion
5. Waiting
6. Unnecessary processes
7. Scrap and rework

1.2   Waste 1: Overproduction

Overproduction

Too Much and Too Soon

You have invited friends and spaghetti is to be served. To make sure everyone gets his fill, you cook the large package. But you end up drinking more than you ate, and twelve portions remain in the pot. This waste due to overproduction is particularly unfavorable, as it automatically leads to further waste. You now must fill the leftover spaghetti in containers, store them in the refrigerator and remember that this stock exists. Over the next few days, you can still use five portions as microwave lunches; the rest will spoil and end up in the trash.

It Is Still Waste

Whenever you produce something earlier or in greater quantity than it is needed at that time, it is overproduction and therefore waste.

You may ask yourself if to produce more is considerate *always* as waste? After all, it does not pay off if I pick three strawberries for each slice of bread, cut them and then boil them with sugar for an hour. Isn't it more efficient to produce a whole jar for a jam, even if that would be overproduction?

If you discover waste in a process, the first step is to identify it. It may well be that no solution initially comes to mind to make the process leaner. However, the ruthless declaration of waste is intended to encourage you to look for and discuss other methods.

1.3   Waste 2: Stock

Stock

Stock Gives a Sense of Security

You are a squirrel and wisely stock up on nuts. In this way, you ensure your calorie supply, even if the supplier "hazel bush" will not deliver over the winter.

But this stock needs space. You must bury the nuts and remember where they are buried. When you retrieve the goods, you first must find the storage position of the nut again. In addition, when the food is finally found and dug up, you find that the expensive commodity is now rancid or completely rotten.

While warehouses provide a reassuring sense of secure supply, they mask many problems. Here, for example, is the problem of unsteady and uncertain delivery. The "hazel bush" simply must deliver more reliably – evolution still has some work to do there!

The Money Is Not Only Fixed in the Goods!

When screws, motors or foodstuffs are stored, money was spent to buy these goods that could have been used more sensibly than left lying around. However, the tied capital is usually not even the most expensive thing in the inventory.

You should not underestimate the expenses for the construction and operation of the necessary warehouse: You need shelves, pallets, bins. Forklifts drive around and computer programs manage content and storage locations. In addition, there is the effort to store and retrieve the material at the right time. People must be hired and trained for this. We count the material for inventory. In addition, since the whole spectacle cannot take place in the open air, we need a hall. We heat in the winter and cool in the summer. All the trappings make storage far more expensive than one would expect based on inventory value.

We often see companies implementing automate warehousing to get a handle on inventory and reduce costs. Fully automated warehouses are the supposed solutions. But we must disappoint again: Even automated waste remains waste. It is often forgotten that automated retrieval and storage are time-consuming processes that should not be underestimated.

Big Trucks and Changeable Weather

Smaller or larger storage facilities can be found in every household and every business. The restaurant Wetterhorn, in a remote place in the Swiss mountains for example, stores large quantities of beer and lemonade in the cellar. After all, the truck comes only once a month. Moreover, the thirst of the hikers is hard to predict. When it is sunny, they come by the thousands and drink kegs dry. When it rains, only one comes and drinks a cup of tea.

Large delivery units, minimum order quantities, volume discounts or fluctuating customer demand are good reasons for inventories. Nevertheless, from a Lean point of view, inventories are always waste! Even when there are supposedly good reasons for it.

1.4   Waste 3: Transport

Transport

Parts Tourism

Transport is always the consequence when process steps are spatially distant from each other – and transport is waste. The production of a T-shirt is a good illustration of this. Cotton is grown and harvested in the USA and then transported by ship for several weeks to China. There, the thread is spun and transported by truck and is on the road again for several days. The fabric is woven and transported again. Finally, the T-shirt is sewn, packed and travels halfway around the globe, back to USA, where it waits for buyers in the store. As was the case with the other wastes, there is always a reason for the waste.

Whether a Few Meters or Halfway Around the World, Both Are Transport

For T-shirt production, it may be economically advantageous to produce in low-wage countries. Nevertheless, the transports are waste in this process.

Short transports also must be organized and coordinated. These add complexity to the process: What is to be transported? *From* where and *to* where? Who will carry out the transport? By what means? The answer is containers, vehicles, drivers, enclosed documents, computer and tracking systems. You need cranes, forklifts or personnel to load and unload the goods. In short, transportation costs money and the goods for the customer are usually not improved as a result. Thus, a clear case of waste.

1.5   Waste 4: Motion

Motion

Even Short Distances Lead to Espresso

Let us come back to our barista: all paths and movements that he carries out in order to achieve the value creation are waste. Because even necessary movements are waste. Actions that the feet or even the hands cover to grab the coffee cup are waste.

Also important for the evaluation of the distances is how many times a day you must cover this distance. Utensils and ingredients that you use a lot, such as coffee cups, naturally weigh more on the waste scale than water or refill bags of coffee beans, which are used less often.

Many Few Also Make a Lot

At the end, many small steps can add up to a marathon distance. It is two meters from the cupboard to the coffee machine. Forth and back, it is four meters. Over the course of a day, during which about 300 cups of coffee are brewed, that adds up to 1200 m. Over the course of a month, our barista sprints a full marathon. Do we want to calculate that over the course of a year? That is over 400 km

Which is more time-consuming: moving the cupboard one meter to the right or walking from Philadelphia to Boston?

1.6   Waste 5: Waiting

Waiting

If It Takes a Little Longer…

Nobody likes to wait. Who chooses the longest line at the supermarket checkout?

Waiting is also undesirable in the sense of Lean and one of the kinds of waste. Our barista must wait a whole 25 s while the coffee flows into the cup. In addition, the coffee did not even really get better while we were watching.

25 s is a short interval, but over the approximately 300 espressos that are prepared every day, even that adds up to several hours. Moreover, the wait was not relaxing either.

1.7   Waste 6: Unnecessary Processes

Unnecessary Processes

Newsletter and Lost Glasses

Everyday life offers many examples of unnecessary processes. Looking for glasses or keys. Opening and closing cabinet doors in the kitchen. Fiddling with stickers from an apple, deleting a thousand newsletters from an e-mail account, or – always fun – picking needles out of a new shirt (only to forget one, ouch!).

Most of the time, we have become so accustomed to all these unnecessary processes that we no longer even think about their tediousness and waste of time. If we do think about it, the excuse is quickly at hand: "That's just the way it is! There is nothing you can do! I don't have time to change anything now." Even here it is worth the effort to change the process in the long run. So, take a moment and finally unsubscribe from these unnecessary advertising newsletters!

1.8   Waste 7: Scrap and Rework

Scrap and Rework

Does Not Always Succeed the First Time

It does not always succeed at the first time. The phone has rung, the oven has been forgotten and, in the end, it has mutated into a smoker grill. Under certain circumstances, the marble cake can still be reworked and saved with an inch-thick chocolate icing. But if the phone call took a little too long and there is more charcoal than cake stuck in the baking pan, the work will end up in the garbage – and along with the cake, money, time and sweat will also be lost. Fortunately, you can notice a charred cake right away. If the sugar is missing, it is only noticed when you eat it – and in front of all your relatives! How embarrassing!

1.9   See and Understand Waste

Yes, But...

"Yes, but the warehouse is necessary!" or "It's much more efficient to produce in large quantities!" may be going through your mind now.

We are aware that we are challenging you with this uncompromising categorization of waste. Nevertheless, we would like to encourage you to simply name the waste in your process. Even if you do not spontaneously come up with an idea of how you could reduce inventory, shorten waiting times or transport routes, the first thing to do now is to make the waste transparent.

If you dare to name the waste and question existing processes, you will start to see them with completely different eyes.

... Sometimes You Have to Look Closely ...

Not everything that looks like waste is waste. In the case of wine that is stored in the cellar and becomes more valuable as it matures, storage is not a waste in the sense of stocks. After all, value creation is going on here.

Some jobs, for example beach lifeguards, consist *only* of "waiting". While sitting motionless like this, they watch the water and peek for drowning people. If the product is called "safe bathing", this activity is important because it ensures safety, but it is still wasteful in the Lean sense. A process that at first glance seems necessary, such as monitoring a critical piece of equipment on the shop floor, is waste. Thus, inspections are always waste, even if they are necessary and can be argued with quality reasons or with safety aspects.

1.10 Now It Is Your Turn

Sharpen your eye for waste. Try to identify waste when shopping at the supermarket, cooking at home or observing processes in your company. And even if a warehouse seems like a sensible measure at first glance, even if overproduction seems totally efficient – put on your Lean glasses. Just call it waste, regardless of the need.

Once you have trained your sense of waste a bit, you can then move on to the somewhat more complex production processes in your company. Observe and recognize waste and keep your head clear of solution ideas or improvement measures for the time being.

With a simple template, you can capture and document waste. Go to the place where the value creation happens, identify, and classify waste. You should do this "waste walk" in a team – together you can see more. Discussing waste and estimating what benefit eliminating waste at a particular point would have work better together. Discussing a potential benefit sharpens the picture to possibly prioritize the implementation later. Initial improvements can always be implemented immediately. In line with the principle "Sensitize – See – Change" note down the possible measures together, e.g., on a flip chart – do not forget to define (*one*) responsible person and to set a clear deadline.

2

The 9 Principles for Eliminating Waste

Contents

A principle is a proposition or value that is a guide for behavior or evaluation (Wikipedia 2020a)

2.1 With 9 Principles to Ideal Production

If you have decided to live healthily, you must adhere to certain principles. Such principles could be, for example, a healthy diet or regular exercise. Whether you go jogging every day or prefer to ride your bike to work – the

© Springer-Verlag GmbH Germany, part of Springer Nature 2022
R. Hänggi et al., *LEAN Production – Easy and Comprehensive*,
https://doi.org/10.1007/978-3-662-64527-7_2

concrete method or tool – is the consistent consideration to implement the principle.

Waste-free production also requires compliance with principles. In the following chapter, we describe nine Lean principles that lead to waste-free production.

Why These 9 Principles in Particular?

Lean literature often does not distinguish between "principles" and "methods". Many of our projects have shown, however, that it is precisely a separation of principles (design guide for the production) and methods (tool to implement principles) that is important for the understanding and success of Lean. First you need to understand the principle – that is, what is to be achieved. Then you can choose the right method to implement it.

From our experience with Lean change initiatives, we have identified nine principles that are important for implementing Lean processes.

2.2   Principle 1: Pull Principle

Pull Principle

Bake, When the Customer Buys

When do we produce how much? The baker does not know exactly how many bread rolls he will sell tomorrow. He does not want to bake too many, of course. That would be overproduction and therefore waste. But on the other hand, he also does not want to alienate any customers by closing time. To meet customer demand as accurately as possible, he has developed a mathematical model that tells him exactly how many rolls he will have to bake for tomorrow, depending on the weather, day of the week and time of year. As he has more and more types of rolls in his assortment, his model becomes more and more sophisticated and complicated, but still the same thing happens again and again: He has not met the exact customer demand. And at the end of the day, unfortunately, a lot of rolls usually ended up in the trash, and yet some customers left the bakery without bread because their favorite was already sold out.

The baker gives up his mathematical model, which tells him how many rolls to push into the market. From now on, he works according to the pull principle. He simply follows a simple rule: "a fixed maximum quantity of the roll stock must not be exceeded". Following this rule, he may only produce if the quantity falls below a defined level under this maximum as a result of a withdrawal (pull). He gets this signal from the empty box on the sales shelf. And according to the rule, he may now also produce only so many rolls that this maximum limit is not exceeded. The defined limit for the baker is the full crate – and not one more roll.

In summary, it can be said that the pull principle is aligned with current customer needs (thought from the customer).

The pull principle is a simple principle that puts the waste "overproduction" within fixed limits. There are many methods to implement this principle. A very well-known one, used by the baker and which we will present in the next chapter, is called Kanban.

Contrary to popular opinion, it is not decisive for the distinction between push and pull whether one produces only for a customer order or whether one produces for stock without a customer order. Nor is essential for the pull principle the technique of information transmission (central, decentral, digital or in paper form). The only thing that is important for the pull principle is that the defined maximum quantity is not exceeded.

2.3 Principle 2: Flow Principle

Flow Principle

Your Name in Your Color

Tim's new business idea "your name in your color" becomes the hit at the weekly market. He has made all the preparations and uses his new truck to transport the freshly produced goods to the market.

"Customized" is his motto, and to that end, Tim researched the 30 most common names and added the most popular colors to the range.

The problem: with the selection of the 30 most common names, he can only serve 20% of the clientele. 80% of those looking for a mug cannot find one with their name on it at Tim's – and certainly not in their favorite color.

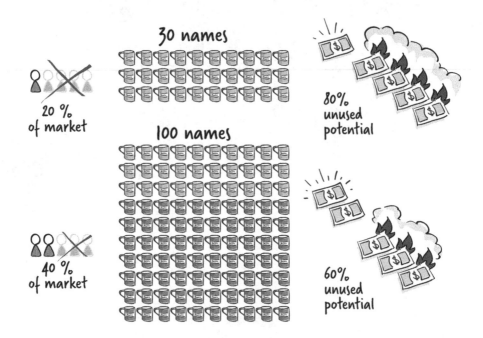

Tim analyzes the situation and realizes he has a conflict: If he were to expand his range to 100 different names, he could already cover 40% of customers' names and double his sales. But his truck is already filled to the ceiling with the top 30 names. He does not have enough money for a bigger truck now, and he does not have any parking space at home either. And Henry Gustav, in 101st place, still would not have a mug.

The cause of the problem has a name: batch production. And so does the solution: the flow principle.

Every time the color is changed, the brush must be washed out and Tim has to search for the right stamp for every name. This costs time and money. All the work involved in preparing production from one product to the next causes setup costs. And that is why Tim always produces several mugs of the same type one after the other, in so-called batches.

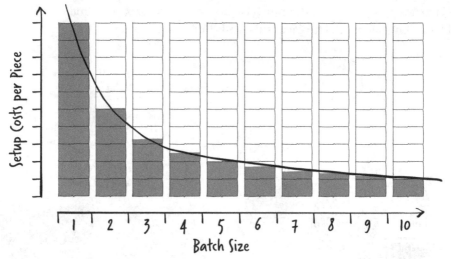

The relationship between part costs and batch size is obvious. If Tim can distribute the setup costs of the special variant over several mugs, the costs per piece decrease. With a batch size of 1 (in Lean language this is called "one-piece flow"), he would have the highest setup costs per cup.

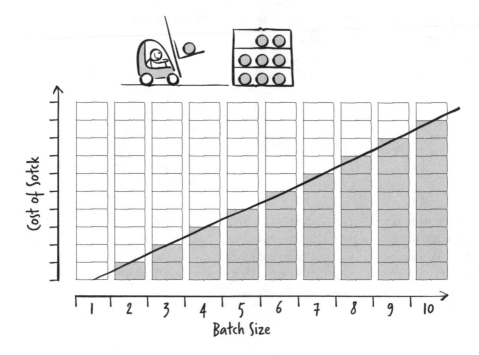

But production in batches also has a downside: Tim causes overproduction because he produces his cups earlier and in larger quantities than customers ask for. And where the mother of waste, the overproduction is, you do not have to look far for her six waste children: stock, transport, motion, waiting, unnecessary processes and the one or other crack in the mug due to the stacking of the batch.

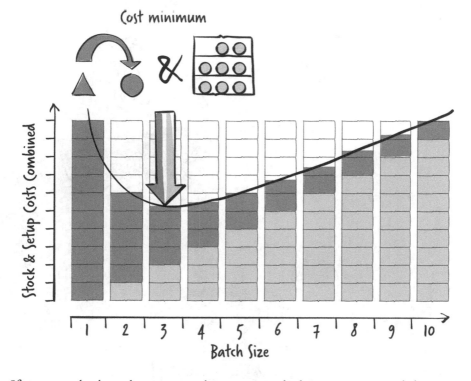

If you now look at the two cost drivers in total: the setup costs and the waste costs due to batches, a certain number of mugs theoretically results in a cost minimum. But even with this optimal batch size, Tim still needs stock – and still does not satisfy all customers. This is not in line with our Lean philosophy – after all, we want to satisfy *all* customers, and ideally do so without any waste.

What can Tim do now to eliminate waste, but still stay at optimum cost and produce?

Tim has (only) one way out of this dilemma: He must tackle the setup costs. His vision: If he manages to push the setup costs for a single copy down to 0, he could also produce the mugs economically in an one-piece flow.

That is why Tim invests in the Mug Master 3000! Instead of a stamp, the mugs are written on using a washing machine-proof plotter pen. And instead of four colors, there are 3.2 million different shades to choose from. The setup costs to switch from Anna in orange to Zoe in purple are practically zero.

Tim now only needs 100 white mugs, and for these, he no longer needs a truck. Inexpensively, with bicycle and trailer, he now travels to the market. And every customer, even if his name is John Dawson, now gets the mug with exactly his name, in his very favorite color.

The opposite of production in batches is our second principle, the flow principle. Tim has now realized a new production concept based on this principle through the Mug Master 3000. He was even able to implement the most radical form of the flow principle: the production in "one-piece flow".

2.4   Principle 3: Takt Principle

Takt Principle

Burger & Co

Every 30 s, a customer comes to Burger & Co. The grill master can easily keep up. However, the second employee needs 40 s to assemble the burgers and is therefore 10 s behind with each burger. Therefore, the fried patties pile up in front of him, creating stocks and thus wastage in the process. The subsequent packer is bored because he only needs 20 s per burger. And this waiting is, again, waste.

You see, the overall process is not "balanced". The customer cannot be served every 30 s and the queue gets longer and longer.

Distribute Content Equally

In order to implement the takt principle, all stations must be oriented to the frequency of the customer orders. In this case, it is 30 s. Therefore, we will shift 10 s of work from the second employee to balance the process. The burger maker is relieved, and the bored packer will now additionally apply the ketchup and put on the bread lid. Now the process is balanced, and everyone has exactly 30 s of work content. This also eliminates stock, over-production and waiting time. At the end of the day, we have sold 25% more burgers thanks to the takt principle.

Only if all processes follow a uniform cycle can waste be eliminated. And if we do not want to build up stock or keep the customer waiting, the takt corresponds to the customer cycle. The takt principle aligns all work processes with the customer's demand. Thus, a flow is generated between the work processes.

The flow principle and takt principle form a logical unit. Takt brings flow and flow simplifies the takt. That is why the takt principle and the flow principle are confused in practice sometimes.

2.5 Principle 4: 0-Defect Principle

0-Defect Principle

Break the Defect Chain

Remember the waste called "scrap and rework"? If you want to get them out of your production processes, you must break the error chain. You cannot accept errors, you cannot make errors, and you cannot pass on errors. This definition of the 0-defect principle sounds logical and simple at first. But if you want to implement it seriously, it requires a radical rethinking. After all, it is easier to order a few more parts – just in case. And who has the time in the operational hectic to analyze the origin of the error in order to eliminate it by its root cause?

A Sustainable Error Culture

The difficulty is that in many areas of life and minds, a less efficient problem-solving process is already firmly entrenched: Find someone to blame and punish him for the mistake – error prevention through deterrence.

However, such a culture is problematic for sustainable troubleshooting in production processes (and other areas as well). The error is not admitted; it is played down and covered up. And that cannot be the basis for the 0-defect principle. You see, the implementation of 0-defects is difficult because it often requires a cultural change.

0-Defect Is Also a Question of Technology

No less important for the 0-defect principle are technical means to detect an error. The technology must help to clarify questions: when, where, how often and under what conditions does the error occur? Only by including numbers, data and facts can a sensible solution be worked out. If you want to introduce the 0-defect principle, this can therefore only be achieved through a combination of technical and organizational measures.

Already at Toyota, in the Lean origins, the implementation of the 0-defect principle was of central importance. Jidoka is the often forgotten and underestimated Lean principle from the Toyota production system. In Taiichi Ohno's original concept, Jidoka, along with just-in-time, is even one of two central pillars of the production system. Ohno understood Jidoka as the separation of human work from machine work in order to achieve 0 defects. The principle is to endow the machines with a humanity or intelligence. For example, the machine can respond to an error by stopping to contain the consequences of the error. This can immediately initiate a root cause analysis and correct the error.

In the origins of Lean in 1950 s, Jidoka solutions were intelligent mechanical solutions, for example, to shut down equipment in the event of a fault. The technical progress of recent years opens completely new possibilities for this idea through "intelligent automation".

Even then, Toyota had realized that the Jidoka idea, the combination of man and machine and the immediate reaction to errors, was starting the road to (complete) automation. Some examples: lane assist warns in case you unintentionally leave the track. The minimum distance to the vehicle in front is undercut and the driver is warned. The navigation system kindly asks to turn around (if possible) if we miss an exit.

These are all mechanisms for avoiding errors, but they are also steps toward a fully automated system. In this case, for example, the driverless vehicle.

We see, the machine must become intelligent for the implementation of the 0-defect principle and recognize errors immediately. This is machine learning at its purest.

The 0-defect principle shows the way. Especially in the times of Industry 4.0, this is more important than ever – Lean and digitalization complement each other. In a 0-defect culture, digitalization leads to the expected savings.

The Four Stages of Intervention

We see four levels in the intelligence of the machine to support the 0-defect principle.

- Level 1: the machine records data to enable subsequent fault analysis. Example: the black box in the airplane.
- Level 2: the machine warns the human before an error occurs. Example: the refrigerator beeps when it is left open for a longer time.
- Level 3: the machine stops when an error is detected. Example: the copier stops when a paper jam occurs, before the jammed sheet causes a larger damage.
- Level 4: machine corrects the error automatically. Example: the counter steering of the lane assistant in the vehicle.

0-Defect Culture

Unfortunately, even the best technology has no value if your organization does not use it properly. To get to the point: the black box in the airplane is of no use if it is not expertly analyzed after an accident and the right conclusions and measures are derived from the findings (Syed 2015).

You must therefore create a parallel organization that guarantees a neutral analysis of the error. All those involved must be integrated into this process without upsetting the people directly involved in the error.

Ultimately, the 0-defect principle means nothing other than learning from mistakes. And of course, you can also learn from the mistakes of others – comparing and exchanging ideas with other companies is therefore important for the 0-defect culture.

2.6   Principle 5: Separation of Waste and Value Creation

Separation Waste and Value Creation

In the Operating Room, the "Surgeon–Nurse" Principle Applies

In a complicated surgery, every second counts. If the surgeon (customer benefit = value creation) needs a scalpel, fortunately he will not have to go to the storage to look for one. The assistant (motion = waste) will give him one at exactly the right time. Here, the processes of value creation and waste are fortunately separated. But even in production, it makes sense to first separate waste from value creation. And this for several reasons: You can eliminate waste better if it is not scattered in several process steps. For example, material transports (waste) can be carried out much more efficiently if they are combined in one route.

From the value creation side (so here, for example, from the surgeon's point of view), errors are avoided that would result from interrupting the value creating activity.

Finally, another, no less important argument to separate value creation form waste: The time for value creation becomes shorter, more constant and easier to plan if it is not interrupted by wasted time.

2.7 Principle 6: FIFO Principle

FIFO Principle

First Come, First Served

A bowl full of apples is an eye-catcher. But unfortunately, the design of these bowls means that the first apple poured in remains at the bottom of the container for an indeterminate period. The FINO (first-in-never-out) principle probably applies here. This would certainly be a source of rejects and search times for rotten apples (waste).

The fact that the first-in-first-out, or FIFO principle for short, is one of our Lean principles is not only due to uncontrolled ripening processes and the associated rejects. Constant throughput times (takt) and always the same pick point (motion) are only possible by adhering to the FIFO principle.

2.8   Principle 7: Minimum Distance

Minimum Distance

Many Roads Lead to Waste

It is probably in the human nature to take the shortest route. Unfortunately, it is not in his nature to build it. The consequence is that, from our experience, the waste caused by unnecessary motion is one of the biggest time wasters in many processes.

When designing the processes, it is therefore important for you to consciously and "on principle" consider how you can keep the motion in the process as short as possible. This applies to walking distances, but also to reach distances and ultimately to every movement. So, the bath towel next to the tub is certainly better in terms of minimal paths, and the cleaning effort in the bathroom (unnecessary process) also decreases. Double reduction of waste.

2.9 Principle 8: Value Stream Orientation

Value Stream Orientation

Everything Has Its Value

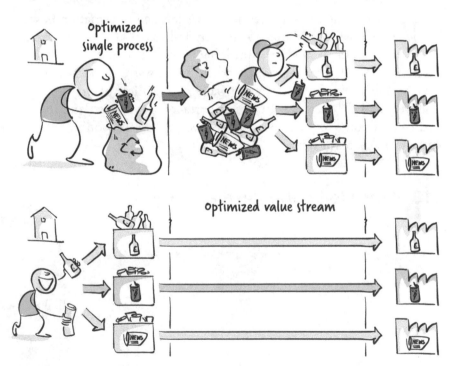

The value stream orientation consists of looking not only at an isolated process, but at the entire chain of processes.

Let us look at the process of waste disposal, for example. Here it becomes particularly clear. For the individual process, it is certainly easier to dispose of everything unsorted. From the point of view of the individual process, separating the garbage would be wasted time. The problem is that separating the waste two process steps further is disproportionately more costly. Therefore, it is important to keep in mind the overall effect whenever a change is made.

The optimization of the value stream takes place "line back", e.g., from the customer to the supplier. In this way, you take on the perspective of the customer.

The sequence of all processes from the customer's order to the delivery of his ordered product results in the value stream. And as our example has shown, you must see the change in processes and elimination of waste in the overall context to assess its effect.

When considering the value stream, the associated information flow must also be considered in addition to the physical material flow. For example, in the case of waste disposal, the information on when and where the waste is collected is central to the design of the entire waste disposal process. Be it in the household, but also in the waste incineration plant, the schedule of delivery and the daily quantity is important. If in Switzerland the school children only come to pick up the paper every 12 weeks, you might want to plan enough space for that.

2.10 Principle 9: Standardization

Standardization

The Lean Anchor

We know the problem from everyday life. We look for our glasses because we constantly put them somewhere else or forget that they are hanging on our forehead. Or we store five different cell phone chargers in the drawer because the plug has been changed for each new device. We can bring up endless examples of how setting standards can prevent waste. The fact that we declare it to be one of our nine principles is intended to underscore the importance and significance for implementing Lean Management that we believe it has.

In continuous improvement, you will constantly develop solutions to eliminate waste. You and your team, with whom you implement the improvements, will invest a lot of time and energy until the solutions fit the specific characteristics of your company and your processes. Then, we have all experienced that the improvements developed with a lot of energy are forgotten after a short time. This makes it necessary to anchor the optimization in the form of standards. Standards can be trainings, process descriptions or even technical tools. Developing and following standards is a core element in building a Lean organization, because developing the solution repeatedly new would be an unnecessary process that would slow down improvements.

It is important to anchor standards in such a way that they serve to prevent errors and make work easier and are not seen as a burden or merely as a control. Therefore, standardize only tried and tested solutions that have been accepted by the entire user community in your company.

2.11 Now It Is Your Turn

We have now presented the 9 Lean principles. Just as with identifying the types of waste, only practice makes it perfect. And you cannot compensate practicing with theory.

Here, too, you can observe which principles are adhered to in everyday processes, for example when shopping or traveling. Or the other way around, where does the absence of principles lead to waste?

Engage with the 9 Lean principles daily and you will better understand how to address waste in processes.

Review some examples of waste in your company that you observed during the last exercise "waste analysis". Now try to identify which of the 9 Lean principles were violated and led to the waste you observed. The answer to this is also the answer to the question: Which of the 9 principles could have prevented waste?

3

Lean Methods to See the Waste

Contents

Identifying the problem is more important than identifying the solution, because accurately presenting the problem leads to the solution.
   Albert Einstein

3.1   Caution: Methods!

You know which symptoms indicate waste e.g., stock, workers walking around or waiting. The link between waste and principles has also become clear to you. Without principles such as tact, flow or standards, lots of waste

© Springer-Verlag GmbH Germany, part of Springer Nature 2022
R. Hänggi et al., *LEAN Production – Easy and Comprehensive*,
https://doi.org/10.1007/978-3-662-64527-7_3

is the consequence. Which tools can now be used to measure wasteful movements or how can you bring tact into the process?

Since the beginnings of Lean Production, science and companies have developed a wealth of Lean methods that can help you implement Lean principles. We will introduce you to the best-known and proven ones on the following pages.

Methods are helpful and important for the successful implementation of Lean. They bring transparency and structure to the change process and help you to see and eliminate waste. Without the systematic application of methods, you will hardly succeed in optimizing according to the 9 principles. But when using methods, caution and a critical eye are also required. Therefore, understand the methods described here more as a suggestion than as a strict instruction. We would like to advise you not to work through them too meticulously. Adapt them to your individual case, only then you will get their maximum benefit. Stay creative and flexible in your search for waste.

These methods are not a complete toolbox. Look left, look right and find your own inspiration for optimizing your production. Even a simple standard report from the IT system or a good conversation with a worker can open your eyes to waste or reveal a solution.

But enough theory and good advice for now. The vacuum cleaner manufacturer LeanClean Inc. urgently needs your help!

3.2   Welcome to LeanClean Inc.

LeanClean Inc. Needs You

LeanClean Inc. is a vacuum cleaner factory, a company with great products and a somewhat dusty production that has grown organically over the last 30 years. Here and there, they have pragmatically, improved the process flow of a production line and once optimized a workplace. Always with great commitment, but with little system or strategy. After the generation change, a fresh wind is blowing in the LeanClean executive suite. And the management really wants to take the "Lean" in the name a bit more seriously now. That is why they also advertised a position to introduce Lean to LeanClean Inc. They were looking for a person who would carry the Lean spirit into the company, live the Lean philosophy and redesign production according to the Lean principles.

You applied and lo and behold, you scored! Congratulations, you are hired! Can we introduce you to our great products first?

The Three Products of LeanClean Inc.

Snake

Positioning:	Cheap line
Quantity:	150 000 pieces/year
Colors:	■
Variants:	None, only one model
Trends:	High cost pressure

Geared to specific customer needs, LeanClean Inc. produces vacuum cleaners in three product lines. The best-selling of the LeanClean products is the entry-level model Snake, a solid standard vacuum cleaner, easy to use and, above all, with an incredibly attractive price. It is not prestigious, but as a cash cow, it fills the coffers of LeanClean.

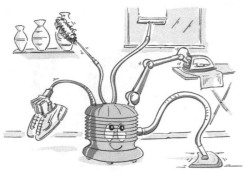

Octopus

Positioning:	All-in-one solution
Quantity:	20 000 pieces/year
Colors:	■ (More colors soon)
Variants:	Additional features in planning
Trends:	New models and options every six months

The Octopus was designed as an all-rounder. Besides vacuuming, it can, for example, iron shirts, wipe windows or tidy up. You can do almost everything with this vacuum cleaner. With its five suction programs, it solves every dirt problem. Now, the Octopus is only available in orange, but the device is soon to delight customers in many other shades. Of course, all this comes at a price, which nevertheless does not deter the large fan community from buying it. To keep these technology-loving early adopters happy, more groundbreaking additional features are planned for next year. These are still secret now. We will be able to tell you more in the chapter "Sequencing".

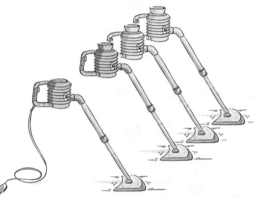

Elephant

Positioning:	Comfort
Quantity:	90 000 pieces/year
Colors:	■ ■ ▢
Variants:	Cable or battery
Trends:	Battery version with increased demand

Finally, we come to the Elephant, the latest innovation in the portfolio. The LeanClean employees produce it in three colors. The device is also available as a rechargeable battery and a corded version. The orange version with battery drive is the best-selling variant.

The Elephant is designed for maximum suction power with minimum weight. It is so light that you do not even realize you are holding a vacuum cleaner. Users report experiencing a feeling of weightlessness while vacuuming. The device owes its light weight to the BlackHole-Generator technology. It is a process that eliminates the hassle of changing dust bags or the icky emptying of dust containers. The integrated BlackHole-Generator simply attracts the dust, as in a black hole, and reduces it to nothing. We will spare you the quantum physics behind this process. After all, in this book we do not want to show you how the vacuum cleaner works, but how to produce according to Lean principles.

The success of the Elephant has exceeded the boldest forecasts: Since its market launch, sales have risen from 50,000 in the first year to 90,000 in the second. Orange is the best-seller, averaging just under 200 units per day. The blue and green Elephant follow with around 100 units per day.

Doubling the output was and is a major challenge for production. Now the masterpiece has also been named "Vacuum Cleaner of the Year" by the Vacuum Cleaner Association. In "Household Gadget Magazine", it made it to the test winner this year. And on top of that, the product won the coveted Infinity Design Award.

Vacuum Cleaner
of the Year
(Vacuum Cleaner Association)

Household Gadget Magazine:
Test winner category
„Battery vacuum cleaner"

Infinity Design Award
„Best Product"

The marketing department and the sales team are celebrating the market success, while the production department is struggling with the increased quantities in addition to the high-cost pressure.

While customers praise the performance of the device, the long delivery times and delays put buyers to a real test of patience. Maybe you have an idea how we can create excitement not only on the technical side, but also on the scheduling side?

A Factory Tour

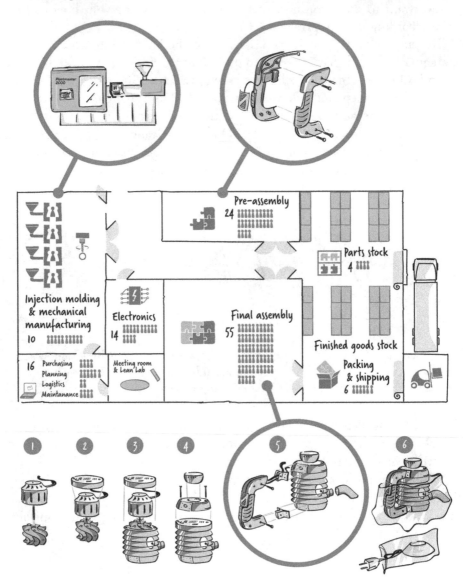

A tour gives you a first insight into the production processes at LeanClean. Unfortunately, you will have to hand in your cell phone now, photography is prohibited, and you will also have to keep the secret of BlackHole-Generator technology to yourself. The three vacuum cleaner models and their individual components are manufactured in the LeanClean production plant. In the injection molding department with four machines, the plastic molded parts are produced. The electronics department supplies the battery and the control system. In the pre-assembly department, components like the handle shells and the switch are put together to modules. And finally, all internally produced components as well as dozens of purchased parts find their way to final assembly, where they are screwed together in five steps and then packed for shipment. 220 employees work at LeanClean, 129 of whom are involved in production.

Three areas will cost you a lot of nerves in your future work. In the injection molding department, the injection molding machines are waiting for you with their erratic quality and long setup times. In the pre-assembly department, the supply of parts is unreliable and delays the delivery date of the entire device with a domino effect. And in final assembly, the takt principle is pretty much out of whack. While the worker from step 5 is drowning in work, the worker from step 2 is so underchallenged that he still has plenty of time to take care of some overproduction.

In addition you notice in your factory tour is that in the production areas there are piles of boxes and material everywhere. There is no free square meter. Even the generously dimensioned warehouse is filled to the last pallet space with raw material, injection molded parts and assemblies. When you see this flood of material, the waste sirens start to wail in your head! Do not worry, you will soon tackle the problem at its root.

The Competition Does Not Sleep

The sales figures for the Elephant show a steep upward trend. However, the success was not entirely free, because LeanClean paid for the increased market share with painful price reductions. The decision to offer the Elephant in three colors instead of just one increased cost on the production side. As a result, margins were attacked from two fronts and have now been virtually eliminated. Where has the profit gone?

A further increase in the number of units is foreseeable due to the great market success of the Elephant. However, production is already at its limit. Investments in additional halls, storage space and employees are not an option for financial reasons. The development of the BlackHole-Generator and features of the Octopus has swallowed up millions. The war chest is therefore empty. LeanClean must recoup this investment over the next few years. But without margins, this is impossible.

More clouds of smoke are coming up on the horizon. The LeanClean CEO stated at last week's employee orientation that the main Asian competitor is opening a distribution warehouse nearby. This gives the competition much shorter delivery times and directly attacks LeanClean's leading position. Faster delivery times are therefore vital for LeanClean.

The strategy in development is further differentiation. But the high number of new variants is an additional challenge for production: new parts lists, drawings and work plans arrive almost daily. Often, they are incomplete and immature, but the customer orders are already in the house. Close your eyes and get through it.

So how are we going to continue? How do we manage to handle the increased volumes and post healthy margins again? Exactly! LeanClean finally must look at waste to make the turnaround and address the future challenges. The recipe is called Lean Production and the project is now in your hands. Your goal is to go to the root cause of waste and eliminate it by implementing Lean principles. Of course, you will not be doing it alone. The LeanClean project team will actively support you. But where does your mission begin? Where are the greatest potentials? Where do you have to attack first?

If you want to identify waste, it has proven effective in practice to first focus on one product or product family and then gradually include other products in the optimization. Because LeanClean faces the greatest challenges with the Elephant, you and your project team will first take a closer look at this product family. But before you start analyzing the waste, we will treat you to a short Elephant product training.

The Innards of the Elephant

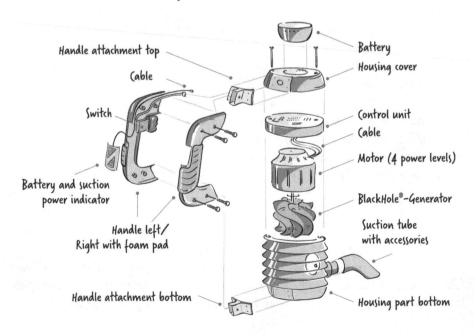

We start our journey with the development view of the Elephant. Later, we will complete the panorama with the view from the production perspective. So here is the Elephant in exploded view with a beautiful view of the BlackHole-Generator and the other jewels of engineering.

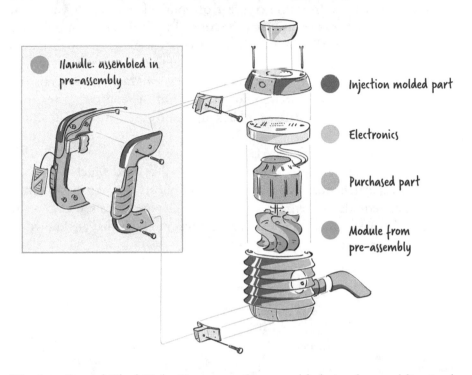

The handle and BlackHole-Generator are assembled as subassemblies in the pre-assembly department. The lid and housing are manufactured internally in the injection molding department. The controller and battery come from the electronics department and the motor is bought in. Now you have a rough overview of which puzzle pieces come together to form an elephant. Then we can start right away with the leanification of production!

3.3 First Analyze, Then Act

Before the surgeon cuts open a patient and inserts a hip joint, an analysis is first necessary. Palpation and X-rays are taken, and a diagnosis is then made. Just as for a surgical intervention, the optimization of a production must always be preceded by an analysis of the actual situation.

You will have to argue for the changes and the best arguments are numbers, data and facts. Ideally, you should collect your own data. Because the principle applies "don't trust any statistics that you haven't faked yourself". Regardless of whether you are a top manager or a project manager, you cannot do an analysis from the comfort of your office chair. If you want to have a say in the matter of waste, the following applies: Go and observe the process in all its reality on site. Count the parts, measure the distances and stop the times. These observations will help you gain the detail and understanding of the process you need. This is not to say that existing metrics or standardized reports do not matter. On the contrary, after your on-site research, they can help you back up your observations with more numbers.

From the abundance of analysis tools, we have selected eight methods that have proven to be helpful in many of our projects and which we always use with great conviction. We tried to use the most common name when naming the methods: Sometimes it was an English term and sometimes the original in Japanese. Therefore, you may discover methods that are known to you but under a different name.

	Over-Production	Stock	Transport	Waiting	Motion	Defects/rework	Un-necessary Processes
Process Map	✓	✓	✓	✓	✓	✓	✓
Value Stream Analysis	✓	✓	✓	✓	✓	✓	✓
OEE	✓			✓		✓	✓
Handling Step Analysis			✓				✓
Operator Balance Chart				✓	✓	✓	✓
Spaghetti Diagram	✓				✓		
Pareto Chart						✓	
Inventory Analysis		✓					

Each method has its own altitude. There are those that help you see the overall process and existing waste as a whole. Others go more into details or a specific type of waste. We start the Elephant's analysis with two methods that aim to look at the production process from a helicopter perspective: "process map" and "value stream analysis". Then we move on to methods that bring to light deeper details of waste.

3.4   Method 1: Process Map

With the process map, you can record processes, visualize them and understand the flow across different departments. Waste becomes visible and transparent.

The Process Map of the Elephant Product Line

So now we will take a closer look at the production of the Elephant. The aim is to understand the manufacturing processes in detail. The process map method is suitable for this purpose. It uses a graphical process representation to describe how the individual steps are carried out in the various departments. This method is something like the primal method for improvement. A process representation, compiled in the team, is the basis for the comprehensive understanding. On this basis, the wastes become visible very easily.

In Lean Management, the principle applies: No change without a deep and networked understanding of the current state. Do you remember the recycling example of the value stream orientation principle? You will only recognize many causes of waste in a linked process if you analyze and understand it as a whole. And this is exactly what the process map method is all about.

In a cross-departmental workshop, your newly appointed LeanClean project team recorded the process map for the production of the switch handle from the injection molding of the handle shells to the installation in the final assembly. After some discussion and agreement on the real flow of the process, the joint result can be seen on the large wall in the meeting room.

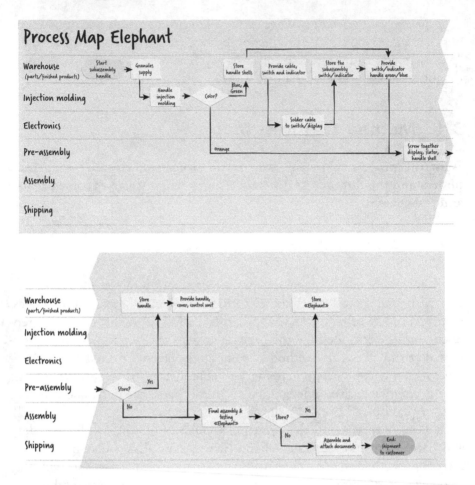

This is how the process for manufacturing of the Elephant handle works: The first process step begins in logistics. In the warehouse, the granulate for the injection molding process of the handle housing is provided and transported to the injection molding department. There, the handle are produced based on a production order and then stored in the warehouse, except for the frequently used orange handle shells.

For the electronics department, cables, switches and the display are outsourced and assembled into the module "switch with display", which immediately end up back in the warehouse until they are needed for in the pre-assembly for the handles.

All necessary parts for the assembly of the handle are assembled for a weekly order and delivered to pre-assembly. Employees assemble the handles, which are then either delivered directly to final assembly or stored.

The workshop participants assumed that all handle assemblies would be stored. Even though direct delivery would make sense for the material flow, there has long been a requirement that every assembled assembly must first go to the central warehouse for proper accounting. Only the discussion has revealed that there is still a direct delivery to the assembly.

The handle and all other parts (lid, control unit, etc.) for the Elephant, if not delivered directly from the pre-assembly, are delivered from the warehouse to the final assembly. The final assembly completes the vacuum cleaner with these parts. Packaged and ready for shipment, they are again put into storage where they await the truck. This is the usual process. But there are exceptions here, too. If products are particularly late, they go from assembly directly to the truck, without a detour via the central warehouse.

Discussion About Waste, Based on the Process Map

You have now reached the first milestone with your project team. The process has been recorded, visualized and understood. Now you and your team need to find out where there is waste in the process. In the first workshop, you identified four places where you noticed a lot of waste. You marked these points in the process map with waste flashes.

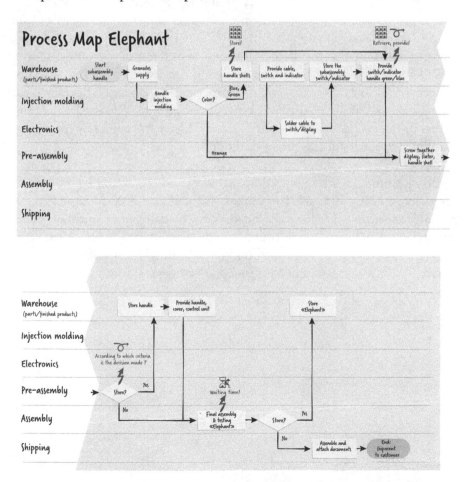

These are the top four wastes of the process:

1. Storage and retrieval of the blue and green handle shells: these process steps are waste in the "unnecessary processes" category. Can't the handle be delivered directly to pre-assembly and save the storage and retrieval? This is already done for the orange handle shell.
2. Storing and retrieving the switch assembly for pre-assembly is wasteful. Here, too, the question arises: couldn't the parts be delivered directly from electronics to pre-assembly instead of taking the detour via the warehouse?
3. It is unclear when the finished handles go directly to assembly or in which case they go to the warehouse beforehand. This distinction leads either to problems with booking or to waste due to additional routes and transport. Either way, the process is not standardized and the criteria for storage are unclear.
4. Waiting times in final assembly are a problem that the participants know well and discussed widely in the workshop. They assigned this waste to the final assembly process. The customer keeps complaining about delivery delays because the completion of the Elephant hangs on this process. Upon closer examination, it turns out that the employee who assembles the handle on the Elephant is overworked and other employees must wait for him. What exactly causes waiting times here, however, requires a deeper look into the final assembly process. The Operator Balance Chart method can reveal the exact reason for the wait time discussed and identified in the process map workshop. You will have to wait a few more pages to learn more about this method, though.

Feedback from all workshop participants was excellent. "After being with the company for over 10 years, I never saw my own process in context", were typical comments. By drawing the process map together, mutual understanding and willingness for process improvement was built. Also, many ideas for improvement opportunities have already been collected. Direct deliveries to the areas, without intermediate storage, are seen by the team as good ways to avoid waste.

Interesting Facts About the Process Map

At the beginning of the process map, there is always the consideration of which process you are analyzing, which departments and areas are involved in the process, as well as which process is being examined exactly. It is important for the result to narrow this down precisely before creating the process map. If you do not realize that this has not been clarified until the recording, it can cost a lot of time.

The process map method consists of two steps. First, you will understand and map the process, and second, you will examine it for waste. The first step, mapping the process as it happens in reality, can be more difficult than it sounds, either because the process was more complicated than expected or because there are different views about how it happens. Either way, engaging with the process is important and even part of the method. That is why it is critical that the process map recording is never done by a single person or only by employees in a single area but is always done cross-functionally and as a team. The creation of the current state in the process map is also important because this is where the first insights into waste and possible ideas for improvement are developed. Every opinion counts! Therefore, use a large area for visualization so that everyone can understand the recording and is included. This way you can make sure that all information and points of view are included in the process recording and that everyone is talking about the same thing. In the second step, a solution is developed that everyone can support.

As you can see from the example of LeanClean, the method is quite simple in terms of content. The processes of each department are entered in the process map in each case in the appropriate sequence in a row. The resulting representation is also called a swim lane diagram because of its similarity to swimming lanes. The process box and the diamond, which represents a case distinction, are the most common symbols in the process map. The flow of material or information from one process to another has been represented as an arrow in our example. For a better overview, you can show the arrows for material or information flows in different colors.

The actual state recorded in this way is an ideal basis for identifying possible waste and pointing it out in the process. If a point of waste already catches the eye during the as-is recording, it can of course also be marked with a flash.

After the complete as-is recording, the team can mentally go through the entire process again with the explicit question of waste. First, it is worthwhile to look at all points where the process changes swim lanes and thus the department. In the example of LeanClean, we had noticed at several such transitions that parts had to be transported and this led to waste. There are storage and retrieval processes at the transitions, which are always waste. You should take a particularly critical look at these points. Transportation, storage and retrieval, and inventory, all are waste. As a team, also check the sequence of machining processes in the same lane. Waste can be hidden here as well. If a process is interrupted, this often results in waiting times or buffer stocks.

Tip

1. Specify the process to be changed and the organization involved. What process will you include? Which departments are involved? What are you excluding from consideration? You should make this delineation before the workshop. Discussions about this in the workshop can unnecessarily eat up time that is then missing for the identification of waste.
2. Involve representatives of all affected departments in the analysis. Especially those who work with the process day by day.
3. Make sure that a moderator guides the recording and discussion. This person must be neutral and should therefore not be an employee of the departments involved.
4. Visualize the process map in a large format. Everyone in the room needs to see what it is all about. For example, use moderation cards or Post-its to depict the processes. Colors help to make the process map even clearer and more transparent.
5. Note problems using waste flashes at the points on the process map where the waste occurs.
6. Always discuss the waste and weaknesses of the process as a team. Never draw the waste flashes alone in the office. The issues are only carried if they have been jointly determined.
7. Use the process map to collaboratively determine the focus issues to work on and to communicate the issues in the process.

3.5   Method 2: Value Stream Analysis

With value stream analysis, you can get a bird's eye view of the whole production process. In accordance with the value stream orientation principle, the aim is to see the big picture and create transparency about material and information flows from the delivery of the raw parts to the shipment of the finished product to the customer (Rother and Shook 2009).

The Value Stream of the Handle of the Elephant

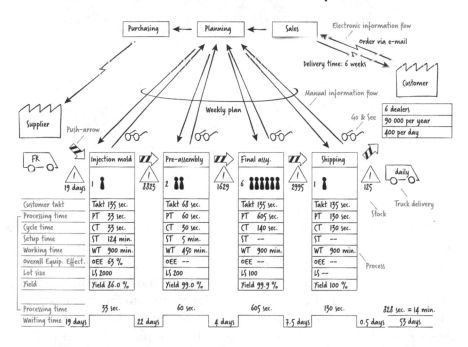

Let us take a look together at the value stream map of the Elephant line that your project team has created. At first glance, the presentation may seem a bit cluttered. But once we have gone through the value stream, the logic and strength of the method quickly become clear. All the key data of the process can be seen on a compact overview. It will then be easy for you to record a value stream yourself and analyze it for waste.

Now we will take a closer look at the four different areas of the value stream.

The Customer and the Supplier

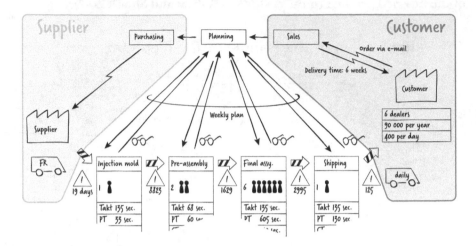

Let us get started and begin on the right at the "Customers" factory symbol. At this point in the value stream, you can find out how many units are required per year and how the orders from the customers reach the production plant. In the customer data box, for example, you see that 90,000 units of the Elephant must be produced and delivered per year. The truck symbol tells you that the units leave the plant in daily deliveries by truck to the customer.

On the other side, on the far left of the value stream, you will see another factory symbol. In this case, it represents the manufacturer of the granules used in the value stream in the first process, injection molding. The truck delivers granules every Friday.

The Flow of Information

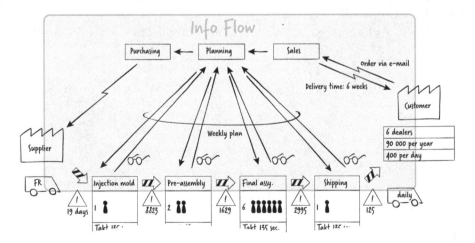

Before we look at the individual production processes and the associated material flow, we consider the information flow. This is outlined in the upper part of the value stream and begins with the order from the Elephants. Orders from wholesalers are first recorded by the internal sales department. A standard delivery time of six weeks, estimated from experience, is communicated to the customer.

Based on current orders, forecasts and current inventories, central planning creates weekly plans for each process. Remember the push principle? Here you can see it in its purest form.

The eyeglass symbols show where there is a manual readjustment within the planning week: Due to unplanned events such as rush orders, sick leave, missing parts, rework or machine breakdowns, there are regular changes in the plan that need to be reacted to.

The production departments report the fulfillment of the orders back to the central control at the end of the week. This can be seen from the arrows pointing from the process back to planning.

The Material Flow

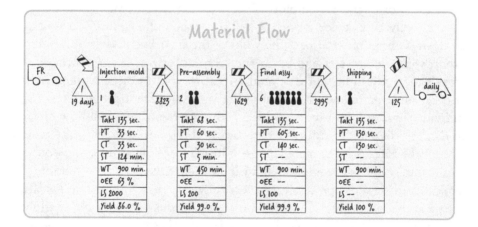

The material flow is the central part of the value stream. We will go through it upstream, backwards as it were, from the customer to the supplier. This way you will take the perspective of each customer of the supply process and better understand their requirements.

The last process in the material flow chain is shipping. All devices to be shipped in the current and next week must have been stored in the finished goods warehouse by then. At the time of recording, 2995 Elephant devices were counted in this inventory.

You can see from the warning triangle and the push arrows that these stocks are controlled according to the push principle.

Shipping

In shipping, one employee per shift works proportionately for our product. He assembles the shipments, prints the delivery papers and helps load the trucks. The processing time (PT) for shipping is 130 s per unit on average.

An important key figure is the customer takt of 135 s in the process box. This is derived from the average number of 400 pieces per day and the working time. The latter is 900 min per day here if the process works in two shifts. All breaks are deducted from this.

After this little digression on the customer takt, let us go back to analyzing the process. Before the shipment is loaded into the truck, it arrives in a buffer area. Here, 125 devices were waiting for the truck at the time of the value stream recording.

Final Assembly

The shipping department is supplied by the final assembly department. The working time (WT) here is also 900 min per day and the customer cycle is 135 s accordingly. Six employees are working here in two shifts and the process delivers a device every 140 s This is the cycle time (CT). But not all devices are in order either. The yield of the final assembly is 99.9% or in other words: 90 devices end up in the scrap every year. To get through the workload of 400 units per day, final assembly workers often work overtime and on some Saturdays. The final assembly of the Elephant requires many individual parts. If you represent the information and material flow of each of them in the value stream, the value stream becomes confusing. You would not see the forest for the material and information flow arrows, and you still would not gain much additional insight from this extra information. To avoid this, it is more practical to select only one part or part family. These are parts that go through the same processes. In this value stream, the lower shell of the already known handle was selected. The findings are transferable for all parts of the same part family. In this case, for all injection molded parts that are assembled in the pre-assembly stage to form a module that is then installed in the device.

Pre-Assembly

Let us move on to the next process: pre-assembly. Here, two employees work in a shift to assemble a handle every 30 s (cycle time CT). In addition to the handles, they also produce other components for the Octopus and Snake. The query of the stock of finished handles in the ERP system says 1124 pieces. However, the team on site has counted 1629. It probably happens that the employees also produce a few handles in advance, past the system. They argue: "We are more flexible this way". You can see how important it is to get a picture on site and to count stocks yourself in order to bring such problems to light.

Injection Molding

The handle shell is produced in one of four injection molding machines. Preparing the machine to produce handle shells takes 124 min (setup time ST). After the machine has been run in and the quality rate is stable, an employee checks up on the machine from time to time and exchanges the full mesh boxes for empty ones. For each order, a batch of 2000 handles is produced, placed in wire mesh boxes and transported to the intermediate storage area. At the time of the value stream map, there were 8823 handles in the different color variants of the Elephant model. 14% of the lower handle shells cannot be used in the injection molding process and must be disposed of. The equipment effectiveness of the injection molding machines (OEE = Overall Equipment Effectiveness) is currently 63%. You will find out exactly what this means in the next chapter. The last station in the material flow is the raw material warehouse. At the time of the value stream map, the inventory for the granules had a range of 19 days across all colors and models.

Now that was a lot of information, but not all of it.

The Timeline

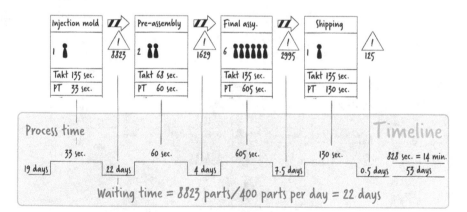

In the lower part of the value stream, a stepped line represents the timeline. You can use it to estimate the throughput time for the entire process. In our case, this is the average time it takes for a granule to pass through the value stream before injection molding and end up as part of an elephant handle in the truck to the end customer. Lead time is an important metric for Lean and a measure of waste in the process.

But how do you interpret this timeline?

Above, the throughput times within the respective process are shown. This is where the value creation takes place. If no parallel processes take place and no large stocks are accumulated in the process, you can estimate the throughput time of the process with the processing time (PT).

Below are plotted the idle times. It is the time that the part or material stacks, waits and lies between processes in inventory, buffers or storage. This is where the waste is hidden. This inventory is given in pieces, but we ultimately want to determine a time. How can we do this? Since each process consumes an average of 400 parts per day of inventory according to customer demand, we can use this assumption to convert pieces into days. We are assuming that the FIFO principle is followed. With this formula, you can also calculate an estimate for the range of coverage in the respective warehouse or buffer.

In summary, the totals are shown on the right of the timeline. At the top, the total of value-added time is 804 s, about 13 min. At the bottom, you see the total of wasted time. A whole 53 days, almost 2 months, the parts are unnecessarily stuck in the buffers and in the warehouse.

Waste Flashes

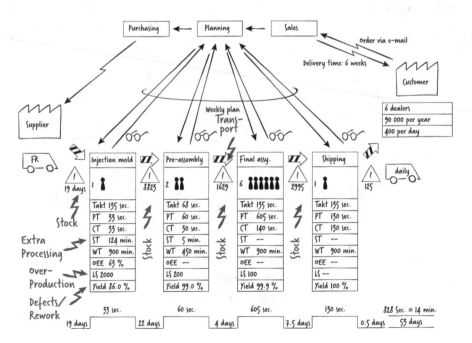

Admittedly, there were a lot of numbers, dates and facts on the last pages. But you do not have to learn the whole collection of numbers by heart. That is why you have the value stream map, which you can consult quickly if necessary. It presents all the details and correlations on one page. And it is not just you who now understands the interlinking and interrelationships in the entire process. Your team now also has a detailed picture of the Elephant's production process.

Now we come to the actual purpose of the value stream map, seeing the waste in the process. We are sure that you have already noticed many places with waste and missing Lean principles in the value stream. We just want to point out some examples of problems in our value stream that lead to waste.

Overproduction
Both during injection molding of the housing parts and during pre-assembly, you can see that overproduction has taken place. The parts are manufactured in batches. This means that the flow principle is missing, and more is produced than the next process immediately requires.

Takt
Final assembly produces an Elephant every 140 s. The customer cycle, however, requires a vacuum cleaner every 135 s. This explains the regular overtime.

Stocks
At each warning triangle, you can see waste caused by inventory between processes. But you can see from the value stream not only *how much* inventory is in each buffer. You can also see *why* it has accumulated there. One main reason is the push control of all processes. Another reason you can see is when large batches are involved, and the flow principle is disturbed. This becomes clear, for example, in injection molding production. Stocks pre-assembled handle modules can also be partly explained by the different shift models between pre-assembly (one shift) and final assembly (two shifts). And finally, we have an inventory of granules that is related to the weekly delivery frequency.

Transport
Transport takes place practically between every process and every transport is a waste. The individual transports of the finished handles by lift truck to the warehouse involve a particularly large amount of waste.

Defects and Rework
The data boxes of the processes show the percentage of good parts (yield) in each process. Scrap and rework are waste, and the lack of the 0-defect principle is certainly a possible reason. Especially in the injection molding process, with only 86% good parts, the problem seems serious.

And Now It Is Your Turn …

The value stream analysis is suitable for providing you and your team with an overall picture. It is a method of taking a bird's eye view of the process and helps to jointly determine which points should be examined more closely. By combining the flow of materials and information and presenting the individual processes with standardized key figures, you can identify where inventories are piling up, where the flow of materials is stagnating, or why waiting times are occurring.

Value stream mapping starts with creating a picture of the entire process chain, from goods receipt to shipping, including information flow. Take a close look at each individual process step and note down all essential data, such as inventories, lot sizes, cycle times or waiting times. According to the go-and-see philosophy, you should go there yourself with your team, see what's going on for yourself, count for yourself and stop the times yourself.

The Elephant's value stream map was done with defined visual symbols that represent processes, inventories or transports, for example. This keeps the representation compact, and everyone can orientate themselves quickly and without major explanations. We have used the common symbols here. Feel free to create your own symbols if you are missing one. How about a forklift or ship transport icon? Just make sure you explain the new icons to your team. This is the only way they will all interpret your visual language in the same way.

After the value stream map and identification of the waste, of course, the work only begins. Together with the team, you must implement measures and methods to eliminate the identified waste.

Tip

1. The value stream map always happens on site and in the team. Go-to-Gemba (which means "go to the place of action") and see together is the motto! Count inventories and stop times as a team. Use data from IT systems only to supplement.
2. First, choose a "racer product", like our Elephant. Value stream mapping is not about mapping the material flow of *all* parts of the product. Ideally, pick a part of the product that goes through as many processes as possible and has a certain value. So do not choose exactly a standard screw (unless you are a screw manufacturer).
3. Choose the part together in the team to increase the acceptance of the results.
4. Everyone in the team needs a clipboard, a stopwatch, A3-size paper, a pencil and an eraser. Now the value stream recording can start.
5. Before your visit, communicate to manufacturing employees that you and your team will show up and record a value stream. Explain that this is to help you to understand the process better. After all, you would think it is strange if a group gathered around your desk unannounced and in silence and started taking notes.
6. Start the value stream map in sales. Ask questions about sales figures, fluctuation and delivery performance.
7. Go to the shipping area and work your way from process to process to the goods receipt of the raw part and purchase of the parts. At each process, you need to understand the most important parameters of the material flow but also the information flow.
8. The knowledge of the employees on site, who experience the process every day, is the most important input for the value stream. No one knows the process better. Ask the employees the right questions and you will get more out of the analysis team than by digging through IT data.
9. If possible, observe the process over several repetitions of the production cycle.
10. Following the value stream recording, it is useful to consolidate the results and record the value stream on a larger format, e.g., on a pin board. Now you can discuss the observed wastes together and mark them with "flashes" at the corresponding point in the value stream.
11. From value stream map to value stream design: once you have become familiar with Lean methods, you can also use the value stream method to draw an ideal target state. What would the Elephant value stream of the future look like?

3.6   Method 3: OEE

What is a metric to see the various wastes and relate them directly to the operational performance of the company? OEE, or Overall Equipment Effectiveness, is that number. It measures the utilization, performance and quality of the process and brings it together.

A Matter of Time Regarding the Plastmaster 2000

LeanClean Inc. uses four somewhat outdated **Plastmaster 2000** injection molding machines to produce plastic parts. During the value stream analysis, we saw that the overall equipment effectiveness (OEE) in the injection molding process is 63%. At first, this does not bode well. Here we show you in detail how the OEE of 63% was calculated, what the important drivers behind the OEE figure are and how you can use the OEE to see waste in the process.

If the **Plastmaster 2000** had produced good parts during the entire target run-time, in the correct cycle, an OEE of 100% would have resulted. Unfortunately, its OEE is currently only 63%, and there are reasons for this.

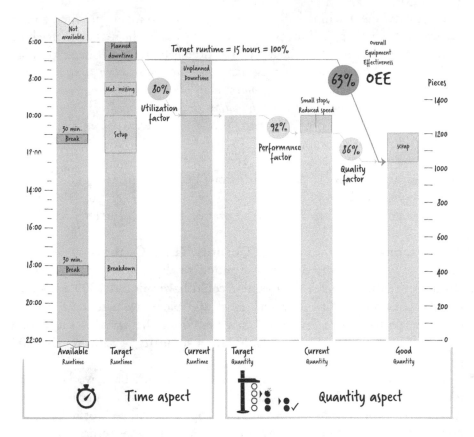

Target Runtime = Planned Runtime − Breaks

First, we determine how long the machine should run at all according to plan at all. In two shifts from 6:00 to 22:00, it is **16 h**. During this time, however, two **breaks** of 30 min each are scheduled. The sum of the breaks of 1 h is **planned downtimes.** This results in a **target runtime of** 15 h. This is the basis for the OEE. Now come the losses.

Actual Runtime = Target Runtime − Unplanned Downtime

In the target runtime, **unplanned downtimes** have led to time losses. What remains is the **actual runtime of** the machine. This is the time during which parts were produced. What losses reduced the result? Once the granulate was missing, later the machine had to be changed over to a new product, and in the evening, another malfunction spoiled production. In total, the machine could not produce for 3 h. The **actual runtime** is therefore reduced to **12 h.**

Note: You will also find OEE calculations that count setup time as one of the planned downtimes. In a narrower sense, downtime during setup is not "unplanned." However, since setup is waste, it makes sense from our point of view to count it also to the OEE losses. It is then a motivation to reduce setup time.

The Utilization Factor = Actual Runtime/Target Runtime

You have produced only 12 h. However, it would have been possible for you to produce parts for 15 h. Thus, the utilization factor is 12 h/15 h = 80%.

The Performance Factor = Actual Quantity/Target Quantity

Now we move from the time consideration to the quantity consideration. With a cycle time of 33 s, the machine could have spit out 1309 parts during the remaining 12 h. However, only 1203 pieces were counted on this day. The **performance factor** is calculated from the ratio of the **actual quantity** to the theoretically possible **target quantity** – 1203 pieces/1309 pieces = 92%.

The performance factor shows you how close you have worked to an ideal output. It can be said that the cycle time represents a maximum, but in practice unforeseen things can always happen, which are not recorded. It may be that the machine ran slower at startup after the disturbances. It may be that there were short stoppages that were not recorded. The performance factor tells you about the loss of volume caused by waste.

Quality Factor = Good Quantity/Actual Quantity

Now, as the last step in OEE, we examine whether the handle shells produced were also usable. How many parts must be reworked or are even scrap? Out of 1203 pieces, we had to sort out 168 faulty parts and only 1035 pieces were OK. This loss is measured in the **quality factor** of the **good quantity in** relation to the actual **quantity** produced. So, 1035 pieces/1203 pieces = 86%. From this ratio, you can see the waste due to defects and rework.

OEE = Utilization Factor * Performance Factor * Quality Factor
For the overall consideration of the OEE, we now multiply the three key figures and obtain the OEE value of the Plastmaster 2000:

OEE = 80% * 92% * 86% = 63%

One can interpret the OEE as the value creation share of the machine time. In only 63% of the planned production time did the machine produce good parts. But the other side of the coin means that 37% or a good third of the time was wasted. At LeanClean, the currently measured OEE has led to great astonishment. "How can this be? After all, our plants run around the clock!". The transparent survey of the OEE has opened our eyes to waste and laid one or two tracks as to how you can avoid it at the Plastmaster 2000.

Caution When Comparing OEEs

OEE is a powerful metric that allows you to condense the effectiveness of a plant or process into a single number. This gives you a quick overview of the effectiveness of different plants, areas even or entire factories. However, comparing the OEEs of different machines and plants is dangerous. This is because OEE is always specifically related to the process, the product and the customer's needs. If a machine runs for weeks with the same product, the OEE, other parameters being equal, is much higher than the OEE of a machine that is retooled ten times a week. You also must consider that for complex products with low tolerances and high-quality requirements, the setup effort can be higher.

Much more meaningful than comparisons between different machines is therefore the observation of the trend of the OEE on the same machine over several weeks and months. In your efforts to be more effective, the OEE tells you whether you are working on the right issues and are successful with your measures. And you can only be successful if you get to the bottom of the causes.

So do not compare OEEs of different products but discuss the three loss factors. This applies to both the machine operator and the production manager.

From Automatic OEE to the Smart Factory

	Use Case OEE	Use Case Predictive Maintenance	Use Case Quality
Extruder Position			●
Injection Pressure			●
Part Weight			●
Pressure Holding Time			●
Temperature		●	●
Humidity		●	●
Mold Wear		●	●
Material Properties		●	●
Machine Running Time	●	●	●
Produced Quantity	●	●	●
Good Parts	●		●
Machine Downtime	●		

Use Case 1: Automatic OEE Recording

Determining and analyzing OEE manually is time-consuming, but can still be useful in the beginning. However, in order to measure the OEE continuously and without great effort, many machines offer the option of reading out the OEE-relevant data and processing it automatically. There are also various software tools on the market that help you to measure OEE automatically. In addition to machine data, data from other IT systems, such as the target cycle time or planned time are necessary to automatically determine the OEE. Therefore, when considering the automatic measurement of OEE, data problems of the existing systems will also come to the table, e.g., because the target cycle times are not correct, and this may have been the case for years. The motto here is – first increase the data quality, and then, you will also get further with the automatic OEE measurement.

The automatic recording of the OEE saves a lot of effort. But it is only a first step, because even if you manage to automatically determine the OEE as a number, the reason for the machine downtime or scrap may still remain hidden. Only when you manage to link the time and duration of the machine downtime with the reason for the downtime can you use the data to correctly interpret the loss and derive the necessary measures. But even here, you can think ahead and create a direct link between context and data automatically. To do this, you need to understand the machine signals in detail, categorize the error codes and assign them to the correct machine signals. Automatically capturing the reasons for downtime, for example by recording machine error codes, helps identify and quantify the causes of waste and improvements in OEE.

Based on the start use case "Measuring OEE", you can build up more complex use cases by including further data, which in turn create additional opportunities to avoid waste. This is a first step on the way to Industry 4.0. Learning from machine data is a central building block here. Two examples:

Use Case 2: Predictive Maintenance
If you could predict the wear of tools, you could also improve the necessary tool change and maintenance strategies of the tools. The tools would be changed and maintained at planned intervals and not fail unplanned in the production process. This would also benefit your OEE. To do this, you need to expand the "automatic OEE" use case to include additional variables such as wear, temperature, moisture and material properties of the tool. Over time, you would see how the variables are related and develop a predictive model for tool wear.

Use Case 3: Quality
The quality rate is an important factor in OEE. The quality of parts on the Plastmaster 2000 depends on many setting parameters. Injection molding is an extremely complex process and adjusting every set screw and machine parameter can have unpredictable consequences for part quality. Undocumented experience is not a reliable guide. If you record machine setting data in addition to the "Predictive Maintenance" use case, over time you will also be able to recognize the connection between part quality and the setting parameters and always set the machine optimally.

If you see automatic OEE recording as a first step toward the digital factory, you can shimmy from one improvement to the next. You can build the necessary IT infrastructure step by step for each use case. Instead of one big investment, one use case finances the next. That is how digitization pays off.

Tip

1. Writing down the times and failures for the OEE calculation by hand is time-consuming and error-prone. Therefore, an IT connection of the machines and an automatic solution for recording the OEE makes sense. There are various software solutions that help here. Use these possibilities. Implementation does not happen overnight, because the technical challenges are often greater than expected. Therefore, proceed step by step. Start with one machine, learn from it and then roll out the concept to other machines.
2. Categorize the reason for failure of the system. The times for failures will not help you if you have not recorded the reason for the failure.
3. OEE also depends on master data. Clean them up first. Then the OEE evaluations will also be correct.
4. Set the OEE target for each machine on a case-by-case basis. OEE should not be a competition in the sense of "who has the best OEE".

3.7   Method 4: Handling Step Analysis

The pure value creation is connected with many processes, such as transporting, testing, unpacking or storing. These are all handling steps and waste. The handling step analysis shows these handling steps in context for a process and helps to see the waste.

MuDa In The Engine

Before your Lean project was launched, a large-scale IT project called MuDa (Monitoring and Universal Data Collection) was planned at LeanClean Inc. The management wanted to use MuDa to create transparency across all processes at LeanClean Inc. For the vision of a central tracking of all handling steps, the management was willing to invest a 200,000 $ budget. In addition, scanner solutions were to be used to ensure seamless monitoring of all processes so that countermeasures could be taken in the event of deviations. The implementation was to start with a pilot project that would initially monitor the entire material flow from goods receipt to the provision of the goods in assembly. The additional effort for scanning also causes ongoing costs, which would require an additional employee estimated across all departments.

After the first Lean analysis, a rethink has taken place. "Don't digitize your mess" is now the motto. After all, not all the waste should be digitized at great expense! Doubts arose as to whether MuDa was the right approach. It was therefore decided to first analyze all processes through Lean glasses and to clarify whether a lot of scanning would even create more waste.

With the help of the handling step analysis, you and the analysis team will bring light into the darkness here. Your focus will be on the process from the delivery of the parts in incoming goods to the provision of the parts in final assembly. As in the value stream analysis, it is best to look at a representative part, for example the motor for the Elephant.

To understand the handling steps in this process, you will go through the path of the motor, from unloading the truck in incoming goods to staging it in assembly, station by station and document each handling step with a photograph. As you do so, make notes about waste, duration or improvement ideas for each step.

The following picture emerged from the handling step analysis for the engine:

Store Parts

1. Waiting on truck
2. Unload truck on buffer area
3. Checking and book goods
4. Bring goods into the warehouse

5. Place pallet in buffer
6. Move mobile rack
7. Pick up pallet
8. Store pallet

Provide Parts

9. Move mobile rack
10. Place pallet in buffer
11. Transport pallet to assembly
12. open packaging

13. Dispose of empties
14. Return

18 Minutes!

The picture shows that an amazing number of steps are necessary before the part arrives at the right place and in the right quantity, in assembly. We go through the path of the motor together from the time it is unloaded from the truck until it is installed in the final assembly. Where do you see waste?

First, the goods must be stored. The truck full of motors drives into the yard of LeanClean. A logistics employee has been waiting for a while on his forklift, ready to unload the goods. Now he drives pallet by pallet to a buffer location. A quality employee makes a visual inspection, and the parts are booked in. Then the motors wait for a ride to their storage location. Then the forklift truck comes around the corner. In the warehouse, too, the motors are first placed on a buffer area before the high-mast forklift takes over the next handling step. Why do we need a high-mast forklift at all? When the warehouse was planned, space efficiency was the top priority, and a six-level mobile racking system was chosen. And in order to reach the higher levels, this kind of forklift is unfortunately necessary. The first thing to do is to wait until the shelves have shifted and the aisle for the selected shelf is accessible. Storing the engine pallet in aisle F, level 4, takes just under 4 min.

A few days later, the logistics department must remove the engines from storage and prepare them for assembly. So, it is the same game, but in reverse order: move the rack, bring down the pallet with the high-mast forklift. In the last step, the forklift brings the pallet to the final assembly line. Now quickly open the cardboard boxes, dispose of them and drive back to the starting point. Done!

What exactly in all these handling steps was waste in the sense of our definition? This question can be answered relatively easily: everything!

Let us quantify the damage to LeanClean together: Each pallet of 50 motors requires 18.2 min of handling. At a logistics hourly rate of 60 $, each engine costs 36 cents more to get it ready for the worker as he needs it. As a reminder, we sell 90,000 Elephant per year. So just for the motor, the handling steps cost 32,400 $ per year. Across all parts of the Elephant, the handling costs add up to a significant cost pool.

The analysis clearly shows how much waste is hidden in the process. And the MuDa project will not reduce these costs by a single cent. On the contrary, MuDa will introduce even more waste into the process and increase the costs of the handling steps. So, it is worth looking at how these handling steps can be shortened or possibly eliminated altogether. It is a good thing that there are methods for avoiding waste. We will come back to the engine handling steps in the "Milkrun" chapter.

What Are Handling Steps?

If you take a close look at your processes from receiving to staging at the point of assembly, you will be surprised how many times the part is unloaded, handled, lifted, sorted, stored, inspected, stacked, picked or transported until it finally arrives where the value creation takes place. All these handling steps are waste from a Lean perspective, and it is worth taking a detailed look at them. With the help of the handling step analysis, you will see which handling steps are made and which enormous costs are buried here.

These are typical handling steps:

- Unloading the truck
- Lifting or setting down the pallet
- Scan, post and acknowledge shipping documents
- Unpacking and repacking containers
- Picking, counting parts
- Moving a mobile shelf
- Transporting the goods
- Storage of the goods
- Removal of the goods from storage
- Labeling containers
- Search/identify the storage location

You will find handling steps in almost all processes. Not only in incoming goods, warehousing or shipping, but also between the value-adding processes in production.

We see a trend that companies are increasingly recording handling steps, process times or quality data digitally by barcode or RFID. They use these technical possibilities to record all kinds of data in any level of detail and on suspicion. In the spirit of a Lean approach, such considerations should always be preceded by an analysis to clarify whether certain handling steps in the process are unavoidable waste. In this way, you can reduce them and dispense with costly digital recording. To ensure that digitization is not used to manage waste, the questions must be clarified in the right order. "Can I make the process simpler and with fewer handling steps?" and only then should you ask yourself, "how can I digitally automate and secure the Lean process"?

Tip

1. Focus the analysis on one representative part.
2. Start at the point of unloading the truck and observe each handling step. Take photographs and notes about the duration of the operation and the process.
3. Calculate the handling costs for the observed part. Extrapolate the costs for the year and for all parts with the same handling steps.
4. Present the results in an overview and discuss them in the team.

3.8   Method 5: Operator Balance Chart

The operator balance chart creates transparency about the distribution of work content between individual process steps. This allows you to see where the process is not running in sync and waste is occurring.

Work Distribution in the Elephant Assembly

During a tour of production, you observe the six workstations in Elephant final assembly for a while. You notice that the packaging employee at the end of the process chain seems a bit stressed and has equipment piling up, while the employee who assembles the engine at station 3 seems to be completely underchallenged.

Something does not seem to be quite right here with the takt principle and the distribution of work content. To identify exactly where the problem is, the team is taking a closer look at the six workstations in the final assembly of the Elephant line. Together with the employees in the area, an operator balance chart (OBC) was recorded for this purpose.

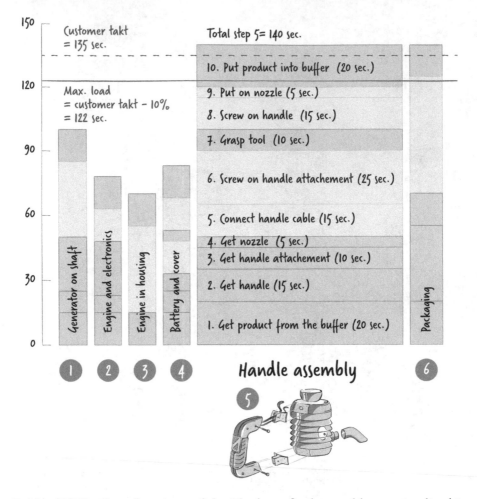

In the OBC, all workstations of the Elephant final assembly are visualized in detail with their work steps.

Each bar in the OBC represents a workstation in the final assembly process. These bars show you the individual work steps, in the processing sequence (from bottom to top) according to their duration. The height of the bar thus corresponds to the respective time in seconds that is required in this workstation for the assembly of an Elephant. The team has shown all work steps whose content is waste in red and the value added in green. You can thus see the waste of each operation. After evaluating the chart, the team used the OBC to come up with three insights:

1. **High Percentage of Waste in the Process Steps**
   You can tell from the small amount of green that the value-added share is less than 50%. In contrast, the larger share is red and is waste. Arguably,

final assembly workers felt the unnecessary walking in their feet every evening, but now it is clearly quantified. Walking time is a quite common cause in the waste analysis. For example, at the handle assembly workstation, steps in which the worker walks account for about 50%.

2. **Unequal Distribution of Work**

   The bars in the OBC are at different heights. From this you can see that apart from the waste in the individual work steps, waste is also caused by a poor distribution of the workload between the workstations. Because there are workstations that take more time to assemble an Elephant than others, the takt principle is violated and overproduction, waiting time and intermediate buffers are created. Waste in its purest form is evident.

3. **Poor Balancing**

   The third insight from the OBC comes from comparing the workload of the individual workstations with the **customer takt.** You can see it drawn in the OBC as a dashed line. As a reminder: the customer takt means that LeanClean must produce an Elephant every 135 s on average so that the customer receives his product at the promised time. Workstations that are below the customer takt produce too quickly. This means that waste is caused by overproduction. Stations that are above the customer takt, on the other hand, tend to produce too slowly. The result: *all* employees in final assembly regularly must work overtime or the customer does not receive his Elephant on time.

   As you can see, the OBC provides a very detailed and multilayered view of the process and its waste. If you take it together with the employees of the area, you will achieve an "aha" effect and gain the willingness to change.

   In the chapter "Line Balancing", you will see how to use the OBC as a basis to distribute work content wisely and thus reduce waste.

Tip

1. The operator balance chart is ideally developed jointly in a team and in a workshop.
2. The more departments that make up the team, the more aspects can be highlighted. However, it is essential that the employees from the affected area of production are represented.
3. Observe the process in the team over several cycles if this is technically feasible.
4. Break down the process into completed sub-processes and give the sub-processes a short, meaningful name, e.g., "fetch nozzle" or "screw on handle".
5. Record the times for each process step with the stopwatch. This time measurement is not a basis for a standard time, according to which work is

done later. The point here is to see waste! Formal time measurements, e.g., according to Methods Time Measurement (MTM) can be used to set up the OBC if they are already available. However, the use of these methods in the workshop is too costly and not necessary for the purpose of the OBC.

6. Discuss the work steps in the team and assess whether the activities are value creation or waste. For this, training on the 7 types of waste and the 9 principles is useful during the workshop.
7. The OBC should be clearly visible to all participants in the workshop so that everyone can understand the process and have their say. Therefore, choose a large projection surface. Depending on the complexity of the assembly and the number of participants in the workshop, a flipchart, a pin board or a larger free wall can be used.
8. Choose a suitable scale for the OBC, e.g., 1 s = 1 cm. Orientate yourself to the customer cycle and the selected format.
9. Build the workstation bars successively from the individual work steps. You can use red and green paper, Post-its or magnetic strips. Cut out each sub-process according to its duration at the chosen scale and build up workstation by workstation so that you get a similar picture to the example in this chapter. Each of these tools has its advantages and disadvantages. Try them out and see what works best for you.
10. Document the collaborative ideas for improvement on a flip chart.

3.9   Method 6: Spaghetti Diagram

You have seen from the OBC that a high proportion of waste is caused by motion. To analyze this in more detail, you can use the simple but informative "spaghetti diagram" method.

A Long Way to the Mounted Handle

With the spaghetti diagram, your analysis team visualized and quantified the waste caused by paths at a workstation of the final assembly. The spaghetti diagram shows the path in the layout that the employee of this workstation has taken within a production cycle.

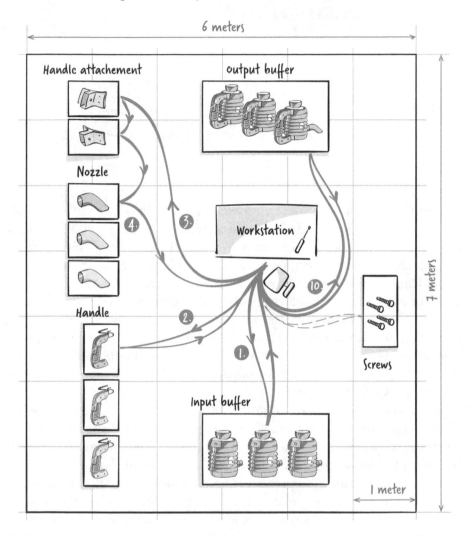

If you take a closer look at the diagram, you can see some interesting numbers, data and facts: per cycle, e.g., for the completion of one elephant each, nearly 33 m are run at this workstation.

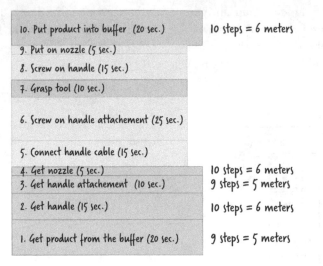

10. Put product into buffer (20 sec.)	10 steps = 6 meters
9. Put on nozzle (5 sec.)	
8. Screw on handle (15 sec.)	
7. Grasp tool (10 sec.)	
6. Screw on handle attachement (25 sec.)	
5. Connect handle cable (15 sec.)	
4. Get nozzle (5 sec.)	10 steps = 6 meters
3. Get handle attachement (10 sec.)	9 steps = 5 meters
2. Get handle (15 sec.)	10 steps = 6 meters
1. Get product from the buffer (20 sec.)	9 steps = 5 meters

The employee walks this route an average of 200 times per shift. In total, he covers 6.6 km per day. In both shifts, 13.2 km. With 90,000 Elephants per year, this makes 2970 km or about the distance from New York to L.A. In the typical product life cycle at LeanClean Inc. of about 4 years, 11,880 km. That is a waste in sense of the seven types of waste, but also a waste of money. If the employee needs one second to cover one meter and the cost per hour in assembly is 70 $, that is 257 $ a day, 57,750 $ a year and 231,000 $ in the Elephant's life cycle. When it comes to financing possible measures, saving 231,000 $ will be a stronger argument than saving 33 m. So try to calculate financial impact. This is true for any methods not only for the spaghetti diagram.

In the above spaghetti diagram of the Elephant, you can also see some of the causes for the many paths. For example, the large mesh boxes with handle attachments, grips and nozzles that can only be moved with a forklift, they use a lot of space and are certainly a cause of the long distances the worker has to go during the process. Is not provisioning in smaller quantities and containers more economical on the bottom line? The arrangement of parts and equipment does not seem ideal. For example, the worker always must make a circuit around his workstation to deliver the finished equipment. Do you see a way to arrange things better?

With the methods for eliminating waste, you can target these points. To do this, you must first identify the waste using the spaghetti diagram.

As you can see, the method is quite simple. Nevertheless, it provides you with an amazing number of details and insights.

> **Tip**
> 1. Explain the method to the employees of the affected workplace.
> 2. Ideally, observe several work cycles first to understand the process and note the sequence of steps.
> 3. On site, sketch out the layout with all the relevant start-up stations such as worktables and shelves on a sheet of paper. It does not have to be to scale to the millimeter. A large step is about 1 m.
> 4. Observe the path the employee takes in the production cycle and draw it on the layout. Count the steps and stop times for one production cycle.
> 5. Calculate the distance covered per production cycle, per day, per year and in the life cycle of the product. Each step is approximately 70 cm.
> 6. Each meter roughly corresponds to 1 s.
> 7. Convert the waste into $ so that it becomes clear.
> 8. Discuss ideas for optimization and calculate the potential.

3.10 Method 7: Pareto Chart

Often the principle applies that 20% of the problems cause 80% of the waste. The Pareto chart reveals the main causes and shows you where you must start in order to eliminate the waste.

From a Gut Feeling to Facts in Two Shifts

Through the OEE analysis, you had found out that the quality rate of the Plastmaster 2000 injection molding machine is only 86%. 14% of the parts end up in the trash. But why are the parts defective?

We ask the machine operator for his assessment. He says that 50% of the parts are out of tolerance. Probably due to tool wear. The adhesion of the soft component, the foam padding on the handle, which is too easy to detach, is perhaps responsible for 20% scrap. And about 10% of the scrap, he says, is caused by an irregular surface, the formation of burrs or a visible flow seam. But is that really the case?

	out of tolerance	Soft component does not hold	Surface is uneven	Burr formation	Flow seam is visible
6.00–7.00	卌 卌	卌 卌 ||	|||	||	|||
7.00–8.00	||	卌		|	|
8.00–9.00	卌 卌 卌	卌 卌	||	|||	
9.00–10.00	|	卌 ||		卌 卌	||
10.00–11.00	|||	卌	|	卌 卌	
11.00–12.00	卌	卌		||	卌
12.00–13.00	||	卌 卌 ||	|||		|
13.00–14.00	卌	卌 卌 卌	|	|||	||
14.00–15.00	卌 卌	|||	|	卌	卌
15.00–16.00	||	卌		卌	|||
16.00–17.00	|	卌 卌		||	||
17.00–18.00		|		|	卌
18.00–19.00	卌	卌 卌	||	|||	卌
19.00–20.00	卌	||		卌 |	||
20.00–21.00	卌 ||	卌 卌 ||	|	卌 卌	|
21.00–22.00		卌 |||		|	|
Total	74	120	14	66	38
Percent	24%	39%	4%	21%	12%

To find this out, we start, as so often, with a white sheet of paper. The machine operator logs the errors here in tabular form in an hourly grid, sorted according to the five types of error mentioned. If a part fails, he marks a line at the corresponding defect type. If a single part has several defects, it is marked several times. The shoe can pinch in several places and so you may end up with more defects than defective parts.

The plant operators put in the extra effort over two shifts with great dedication and curiosity. Thanks in advance. Instead of a gut feeling, we now have real evidence. Let's get down to the analysis. In first place on the defect charts, with 39%, is the insufficiently adhesive soft component, which in this case detaches from the handle. Second and third place go to the tolerance and burr formation.

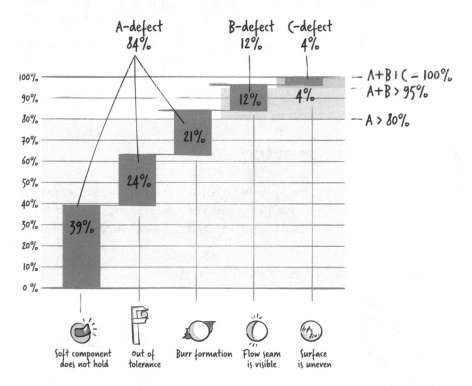

These three causes of errors already account for about 84% of the total errors. And so we have a clear order in which your team will investigate each cause.

When analyzing waste, it is important to get to the bottom of the matter and understand the different causes that lead to waste. Facts are better than opinions. In our example, you have seen that manual recording is usually sufficient for an initial statement. Of course, the support of an IT system would sometimes be desirable, but often the necessary data is not available there or the systematic recording in the computer eats up too much investment and time. As General George S. Patton said: *"A good plan, violently executed now, is better than a perfect plan next week."* Pareto logic can be used universally in many other areas of waste analysis. What are the main causes of inventory, delays or waiting times? A Pareto evaluation will give you a good first answer.

3.11 Method 8: Inventory Analysis

The inventory analysis helps you to illuminate the waste caused by inventories step by step. This gives you the necessary understanding to subsequently reduce inventory with appropriate Lean measures.

360-Degree View of LeanClean Inventories

You can see from the process map and the value stream that inventory has accumulated virtually everywhere along the value chain. This is waste and therefore we will examine these inventories a little more closely. What, and more importantly, how much exactly does LeanClean Inc. store in each of these places? So, let us dive into the details one more time.

The survey of the last inventory says that there are 1185 different parts in LeanClean Inc. And the total value of the inventories is put at $1.4 million. But these two numbers do not really get you and your team anywhere. They only scratch the surface. What will help you is a differentiated view of the inventory.

First, we classify the parts in the categories of cost and consumption. We have divided each dimension into three levels.

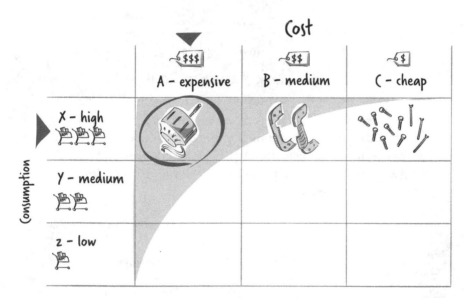

We call expensive parts (A) with a high consumption (X) AX parts. It is worth taking a closer look at this class, as there is the highest potential for waste. There are 72 parts in this AX class, including, for example, the engine that is installed in every Elephant.

It is interesting to ask **"how much"** inventory is available for each of the 72 parts. The number of pieces is not the only valid answer. You can also describe the inventory as a value in $, quantify it as a range in days or express it as a volume in cubic meters.

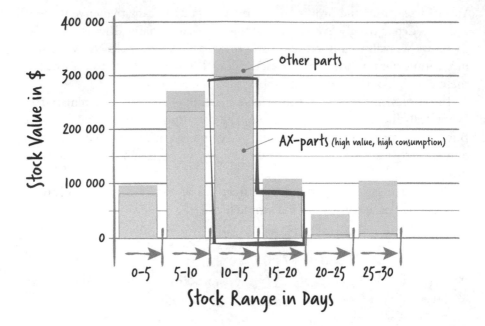

For our consideration of "how much", we order the parts according to the stock range in days and visualize their value in $. To focus on our AX parts, they are shown in a different color. Again we have narrowed down the problem: the AX parts with a large range of coverage and a high inventory value are the focus of your fight against waste.

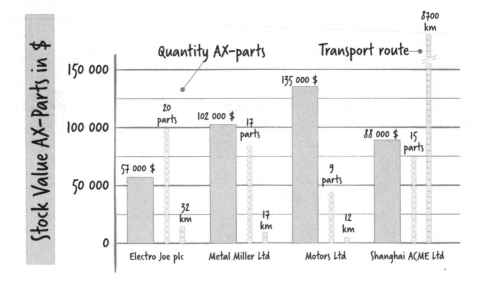

Each further evaluation brings further insights. So, you can divide the parts into purchased parts and internally produced parts. In the case of purchased parts, we are also interested in how far away the suppliers are. Now, we can estimate how fast we can replace the goods. For the AX parts with a range of more than 10 days, we have identified the most important suppliers. You can now see that many of the AX parts that take up valuable space in the brimming warehouse is delivered by three suppliers that are a maximum of half an hour away from LeanClean. The delivery time is a maximum of one day.

You summarize the most important key statements again after the inventory analysis: Only a few of the 1185 part numbers are expensive and have a high consumption (AX parts). These 72 AX parts make up a significant part of the stock value and many of them have a large range.

Most suppliers of AX parts with >10 days range have a transport distance of less than 32 km and can therefore be reached within a day's drive.

This initial analysis has given you and your team a first impression of the inventory and a crucial starting point for inventory reduction. You have seen from this example that in addition to the ABC–XYZ classification, other factors need to be considered in order to get a complete picture of the inventory situation. Thus, you can focus on those parts, which cause a particularly high portion of waste by stocks. The detailed view will help you later to choose the right method to implement the principles and reduce waste.

Tip
1. Evaluations of inventories often have an emotional aspect. That is why the evaluations are always made with the involvement of the various parties involved, such as logistics, planning, production and finance.
2. You may encounter data problems during the evaluation. Agree on a roadmap to fix them.
3. An important aspect is the communication of the inventory analysis. Here it is important not to "accuse" anyone.
4. Set clear goals and responsibilities to reduce inventory. Use the methods to implement the principles to achieve these goals.

4

Methods for Implementing the 9 Principles

Contents

© Springer-Verlag GmbH Germany, part of Springer Nature 2022
R. Hänggi et al., *LEAN Production – Easy and Comprehensive*,
https://doi.org/10.1007/978-3-662-64527-7_4

It is not enough to know; you must apply it. It is not enough to want it, you must do it.

Johann Wolfgang von Goethe

4.1 Only the Implementation Brings the Benefit

You have now thoroughly analyzed LeanClean and know exactly where waste has settled. Now the hard work begins; it is time for implementation. You must now bring in the results. As for the analysis, there are also methods for implementation that have become established in Lean Management. They will help you to change the production process physically and organizationally so that it works according to Lean principles. In addition to methodological knowledge, we also want to give you a feeling in this chapter for the conditions under which the use of a method makes sense and which preparations are necessary so that your efforts lead to the desired result.

We have created an overview of the most important implementation methods for you. In this overview, you can see which principle is focused by each method. Always apply these methods in a targeted manner to a problem that you have discovered in the analysis phase. Using Lean methods only as an end in themselves may lead to change, but usually does not lead to improvement.

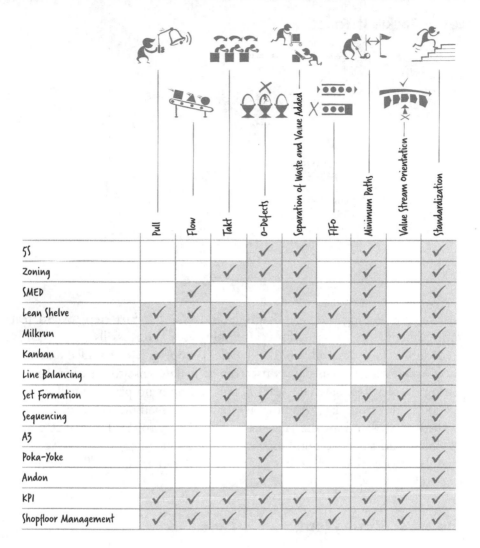

	Pull	Flow	Takt	0-Defects	Separation of Waste and Value Added	FiFo	Minimum Paths	Value Stream orientation	Standardization
5S				✓	✓		✓		✓
Zoning			✓	✓	✓		✓		✓
SMED		✓			✓		✓		✓
Lean Shelve	✓	✓	✓	✓	✓	✓	✓		✓
Milkrun	✓		✓		✓		✓	✓	✓
Kanban	✓	✓	✓	✓	✓	✓	✓	✓	✓
Line Balancing		✓	✓		✓			✓	✓
Set Formation			✓	✓	✓		✓	✓	✓
Sequencing			✓		✓		✓	✓	✓
A3				✓					✓
Poka-Yoke				✓					✓
Andon				✓					✓
KPI	✓	✓	✓	✓	✓	✓	✓	✓	✓
Shopfloor Management	✓	✓	✓	✓	✓	✓	✓	✓	✓

4.2   Method 9: 5S

Stable processes are built on order. The 5S method creates this basis by sorting, making visible, cleaning, standardizing and securing the standards. With 5S, a visible change is achieved and the mindset for Lean is created.

Let us Tackle It Together

We are now leaving the analysis phase and taking a first step toward real improvement. And in the final assembly, your Lean skills are urgently needed. Step 5, the assembly of the handle, is hopelessly overloaded and cannot keep up with the customer's pace. So, it makes sense to start here. Some getting rid of some material, cleaning and setting it up properly will do lots of good to this workplace. And that is what the 5S method is all about.

The five S's stand for the Japanese words Siri, Seiso, Seiton, Seiketsu and Shitsuke. And since we don't assume you're fluent in Japanese, we've used English S-words that may sound a bit awkward, but are hopefully a bit more accessible.

The 1st S: Sorting (out)

After a brief introduction to the topic, we get down to the first "S". Sorting out is the motto. The team has the task of separating everything that is no longer used in the workplace. And *everything* means *everything:* tools, parts, fixtures, supplies and some private items, in every drawer, in every cabinet, in every corner. Employees sorted things into three pallets:

Dispose:	Things that will never be used again
Question mark:	Things whose use is still unclear
Keep:	Things that are needed in this or another workplace

After two hours of hard work and several discussions about the necessity of some items (which have not been used for decades), all three pallets are full, especially the one for disposal.

There are tools in there that have not been used for years or are even broken. There are tools that were borrowed from other departments months ago and are finally finding their way back there. And there are also hundreds of small parts and documents in the pallet that no one knows anymore what they are useful for.

If you encounter larger objects or entire machines in this sorting step that do not fit into the pallet, you can of course also mark them with adhesive dots: red = dispose, yellow = unclear and green = keep.

Together with the team, you have most recently dedicated yourselves to the unclear cases in the middle pallet. After some pain of separation, most of the cases have been disposed as well.

Just by disposing of unnecessary items, the workplace already looks much more clean and organized after this first S.

The 2nd S: Shine

Dust and stains have settled on the work surface over the months. The second step is to clean and sanitize the workplace. All machines, tools, equipment and the floor will also benefit from your care. Besides improving the appearance of the workplace, cleaning also has the effect of making you look at things a little more closely and take the parts in your hand. In this way, you consciously recognize and assess their condition. Your scrutinizing gaze spies a few worn tools again – they, too, find their destination in the disposal pallet.

While cleaning, you and your colleagues already develop a feeling for which points are important for the cleaning and maintenance plan. This inspiration is valuable for the fourth S, standardization.

At the end of the cleaning campaign, the workplace shines and only what is really needed on a regular basis remains. What needs to happen now that employees can find their tools and materials effortlessly and have them quickly at hand in the future?

The 3rd S: Set in Order

It is now your turn to "set in order". Now you assign each tool and each screw box its clear and marked place at the workplace. Hard foam inserts are suitable for the tool trolley. These can be quickly cut by hand and now provide each tool with a suitable recess. In addition, you can label the recesses. This way, each tool is always ready to hand in always the same place. And if a screwdriver is missing, you can notice this immediately.

Making things visible does not just apply to the workplace, but to the entire area with floors, walls, doors and logistics areas. Therefore, grab a wide yellow tape and mark the paths and areas to provide orientation for the logisticians. Draw the lines nice and straight. A little bit of esthetics will greatly increase the acceptance of your work. In the final assembly of the Elephant, we set a new benchmark in terms of visibility for LeanClean. Couldn't other areas benefit from that? And wouldn't it help if, for example, our color codes for floor marking were uniform throughout LeanClean? On to the next step: standardize.

The 4th S: Standardize

The team has put a lot of work into placing, marking areas and labeling each item in the workplace. It would be a waste to work out these solutions anew in each subsequent workshop. Therefore, as part of the fourth S, standardization, you considered which of the solutions worked out so far in the workshop should be declared standard within LeanClean. To record these best practices and make them accessible to everyone in the company, the team created a document called "LeanClean Standards". This now serves as a guide in every workshop and new proved ideas are added here.

In this workshop, the team declared three things about LeanClean standards:

1. **Inserts for the tool trolley:** The development of the tool inserts, the so-called shadow boards, took a lot of work. Hard foam boards first had to be researched and tested. This will go much faster in the next workshop.

2. **Markings:** Markings of logistics areas and travel routes have been uniformly defined. If you are now in the production area at LeanClean, you know where to walk.
3. **Cleaning and maintenance schedules:** Finally, the team agreed on a standard cleaning and maintenance schedule for all workstations and equipment. The workstations are now always clean and uncluttered, and regular maintenance of equipment and tools has measurably reduced the failure rate.

The 5th S: Securing the Standard

The workplaces look like new after the first wave of workshops. Production is unrecognizable. All tool sets are now complete again and employee motivation is noticeably high. But when will the first wrench be missing? Is it possible that everything will revert to the old mode? It's not only possible or likely, but absolutely certain! That is where the fifth S comes in. You want to secure the conquered standard. It is therefore necessary that you check from time to time if the employees keep the agreed standard. If they are not, you must re-establish it together. To assess the 5S level, you have created an audit plan with seven simple but clear questions.

Are not needed tools and
items in the area?

Is waste properly sorted?

Are equipment,
tools and supplies clean ?

Are maintenance and
cleaning schedules being followed ?

Are all tools and parts
in the designated location?

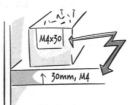

Are all racks, bins, and tools
labeled?

Is personal protective
equipment being used??

A 5S audit now takes place monthly in each department of LeanClean. The result is discussed with the employees of the respective department and measures are derived. It is important that all employees at LeanClean are aware that the standards were created and agreed upon together and that 5S is the basis for a Lean company.

Tip

1. For the first 5S workshop, choose an area where employee motivation and openness to new things is high. It is important for the rollout to have a success story in the first workshop or 5S activity and thus win the organization over to 5S.
2. Start the workshop with an introduction to 5S and recap the 7 types of waste and 9 principles. It is important that employees understand the connection between 5S and waste. If this understanding is missing, 5S will be labeled as just cleaning.
3. Before you start with the first S, do a walk-through in the department with the team. Discuss waste and the connection with 5S. What is the importance of labeled storage locations for routes and transport? What impact has missing labels or markings for search times of tools? What significance does a clean and maintained facility have for the quality of processes and products?
4. If it does not already exist, create a "Book of Standards" to collect the good ideas from all the 5S workshops.
5. After an initial 5S workshop has been conducted in all areas, you can start with the 5S audits. They are crucial for sustainable success.

4.3   Method 10: Zoning

Ways and unnecessary movements are waste. With zoning, all items in the workplace are arranged in defined zones to avoid this kind of waste. Decisive for the zones and the arrangement in the zones is the size, distance and frequency of use of these items.

Subdividing the Workplace

After you and your analysis team recorded the spaghetti diagram, it became clear that the assembly process contains a lot of motion or movement. You now need a way to minimize these paths in the process! To bring even more methodology into this consideration, you and your team divide the work area into three zones and mark these areas in the layout in green, yellow and red.

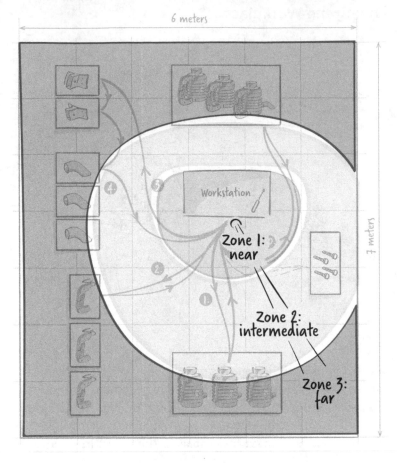

Zone 1 = Near

Here, all the material is arranged within reach. To grab a part or tool within this zone, the employee does not even need to take a step. Therefore, you place items here that are used very often.

Zone 2 = Intermediate

Here, the employee must already stand up and walk two to three steps to get the material or needed tool. Everything that has no longer found a place in zone 1 is arranged here. This material is used less frequently, but still regularly.

Zone 3 = Far

Special tools, rarely used materials or refill containers for screws and small parts are used even less frequently and are not needed in every production cycle. They find their place in the more distant zone 3.

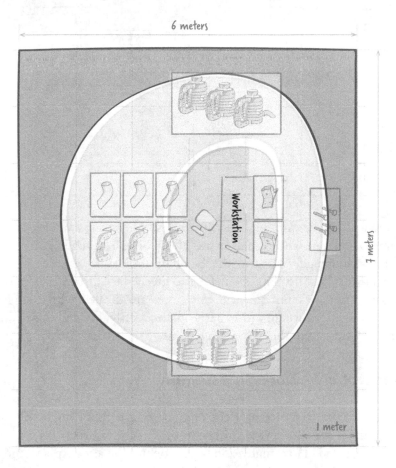

Zoning clearly shows that a different arrangement of material would be more convenient for the employee's walk. One possibility is to arrange the parts not by part groups such as nozzle, handle attachment or handle, but by frequency of use. After all, the orange parts are used much more often than the blue and green ones. Consequently, you will rearrange all the tables, shelves, tools and parts by frequency of use in these zones. The handle attachments that are used in each variation and thus most often get a prominent place in Zone 1, right at the workstation.

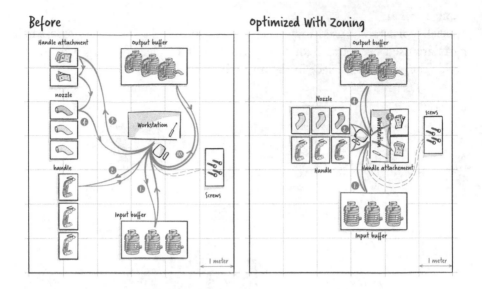

After organizing the working area by zones, you created a new spaghetti diagram to measure and visualize the effect of the change. For the most common, the orange variant of the Elephant, you were able to reduce the distances from 33 m to 8 m. Even for the green slow turner, it is now only 12 m. After initial skepticism, the employees are thrilled with the change. They no longer must carry the parts unnecessarily far and have quickly become accustomed to the new arrangement.

Zoning Not only Works in Assembly

Zoning offers you a simple and effective method to implement the principle of minimal paths. However, there are no precise instructions on how to apply the method. How many zones you define for a workplace and how big they are. All is up to you. It depends very much on the type of workplace and the product. Is it an assembly line in automotive production? Is it a parts warehouse? Is it a workstation where Swiss handmade watches are assembled? Every workplace has completely different requirements. Accordingly, the classifications of zones is also very individual and must be worked out, discussed and defined for each workplace. Here are two examples that show how versatile zoning can be used and how differently the zones are designed in each case:

How can the zones be divided for an office workstation? The mouse, keyboard and coffee mug are in constant use and must be placed in zone 1, while the scissors, hole punch or printer are placed in zone 2.

In the warehouse, zoning is organized according to the frequency of storage operations. Of course, not only routes, but also size and weight criteria of the stored goods must be included in this consideration and the warehouse must be divided into appropriate zones.

> **Tip**
> 1. Follow three steps to organize an area into zones.
>   **Step 1:** Divide the workplace into areas that have equally good conditions in terms of paths. These are your zones.
>   **Step 2:** Prioritize all materials, parts, tools, etc. by importance, such as frequency of use.
>   **Step 3:** Assign the elements from step 2 to the individual zones according to their priority.
> 2. Record a spaghetti diagram before and one after the change. This way you can measure and prove the effect of zoning.
> 3. Space for zoning can be generated by an prior 5S project.
> 4. The zoning concept must be regularly reviewed to ensure that it is up to date. Product changes or changes in sales require a reconsideration of the assignment to the zones.

4.4   Method 11: SMED

SMED is the method with which you will reduce your setup times. After a thorough analysis of the setup process, technical as well as organizational measures are used.

The Plastmaster 2000 Is Aiming for the Pole Position

According to the value stream recording, the mold change for the handle housing on the Plastmaster 2000 takes 124 min. Due to the many different parts produced on this line, the employees must perform many such setups per month. All of these installations eat up valuable capacity on the machine, as it stands still during the most time of the setup. However, to ensure that there is still enough production time left and that the accumulated setup times do not become too large, LeanClean produces its parts in large batches.

Each additional part produced on the Plastmaster 2000 makes this strategy even more problematic. The time until the next part is being produced again becomes even longer as a result, and inventories get even higher. There is only one alternative to get this situation back under control: Setup times must be reduced immediately!

Take Formula 1 as an example: After years of optimization, the teams have managed to change the tires at the pit stop in less than two seconds. If you make it your goal, you will certainly manage to speed up the tool change as well. You just must be methodical, as with all problems. The method you use with your team to reduce setup times is called single-minute exchange of die (SMED).

Like changing a tire in Formula 1, the Plastmaster 2000 changeover process consists of many individual activities. The better you and the team understand this sequence, the more precisely you will be able to say where and how the minutes are wasted during the setup of the Plastmaster 2000. Based on this, you can then define an ideal sequence of individual operations.

At the next mold change, you and your team are on time at the Plastmaster 2000 to take a close look at all the steps of the changeover process.

Recording the Setup Steps of the Plastmaster 2000

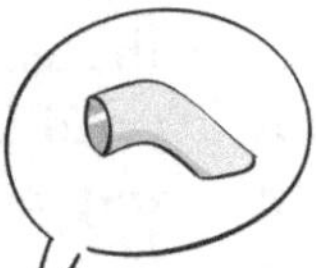

			Time (min)	Int. mandatory?	External possible?
	1	Review work plan	8		
	2	END SERIES "NOZZLE GREEN"	-		
	3	Get granules from the warehouse	10		
	4	Get tool in store	10		
	5	Clean tool	7		
	6	Provide open-end wrenches 8,10,15	3		
	7	Allen wrench missing, borrow	5		
	8	Wait until crane is free	10		
	9	Remove tool with crane	12		
	10	Lift new tool with crane	5		
	11	Position and tighten	15		
	12	Screw on hoses (one defective)	10		
	13	Get replacement hose from store	8		
	14	Mount spare hose	6		
	15	Start up machine	10		
	16	Produce and check test parts	5		
	17	Adjust, produce test parts	5		
	18	Release production in ERP	3		
	19	START SERIES "HANDLE SHELL ORANGE"	-		
	20	Fill out Q-document	7		

Total time required in min

Internal	124
External	15

Left margin labels: Preparation (external); Setting up During Shutdown (internal); Postprocessing (external)

Equipped with paper, pencil and a stopwatch, you are ready to start the breakdown with your team. The machine operator starts the setup process. Even if you now want to pester him with questions, keep a low profile. He should work as undisturbed and as usual as possible. If he were to explain the entire process to you during the recording, your time recording would be distorted. You observe the machine operator and note down what is happening and the time required for each step. It is not at all easy to separate the setup process into logical sub processes when you observe it for the first time. However, this division is important for the method. Like everything in life, recording a setup process requires practice. At the end of the process, you have observed 20 work steps during 139 min. During this time, the line was at a standstill for 124 min. Now that you have recorded the process, it is a matter of shortening the setup time.

Analyze and Organize

		Time (min)	Int. mandatory?	External possible?
Preparation (external)				
1	Review work plan	8		
2	END SERIES "NOZZLE GREEN"	-		
Setting up During Shutdown (internal)				
3	Get granules from the warehouse	10		X
4	Get tool in store	10		X
5	Clean tool	7		X
6	Provide open-end wrenches 8,10,15	3		X
7	Allen wrench missing, borrow	5		X
8	Wait until crane is free	10		X
9	Remove tool with crane	12	X	
10	Lift new tool with crane	5	X	
11	Position and tighten	15	X	
12	Screw on hoses (one defective)	10	X	
13	Get replacement hose from store	8		X
14	Mount spare hose	6		X
15	Start up machine	10	X	
16	Produce and check test parts	5	X	
17	Adjust, produce test parts	5	X	
18	Release production in ERP	3		X
Postprocessing (external)				
19	START SERIES "HANDLE SHELL ORANGE"	-		
20	Fill out Q-document	7		

45 minutes! ←

17 minutes! ←

Total time required in min

Internal	124
External	15

In Formula 1, changing a tire in two seconds is only in advance possible because many activities have already been prepared for the pit stop. No one will slurp into the warehouse to get a tire after the racecar arrives. When the car gets into the stop box, everything is ready and in the perfect place. Everybody in the team knows exactly how each step is to be performed in order to minimize the time of stoppage.

In the workshop, you therefore first clarified together which steps can also be carried out as preparation or post-processing while the machine is running. We call these steps external setup processes. Activities that can only be carried out when the Plastmaster 2000 is at a standstill are called internal setup operations. You now move all setup steps marked as external to the preparation or post-processing phase. This purely organizational change to the work steps has not yet reduced the working time per se. Nevertheless, you have already gained over an hour of Plastmaster production time for each changeover!

			Time (min)	Int. mandatory?	External possible?
Preparation (external)	1	Review work plan	8		
	3	Get granules from the warehouse	10		X
	4	Get tool in store	10		X
	5	Clean tool	7		X
	6	Provide open-end wrenches 8,10,15	3		X
	7	Allen wrench missing, borrow	5		X
	8	Wait until crane is free	10		X
	2	END SERIES "NOZZLE GREEN"	-		
Setting Up (internal)	9	remove tool with crane	12		
	10	Lift new tool with crane	5		
	11	Position and tighten	15		
	12	Screw on hoses (one defective)	10		
	15	Start up machine	10		
	16	Produce and check test parts	5		
	17	Adjust, produce test parts	5		
Postprocessing (external)	19	START SERIES "HANDLE SHELL ORANGE"	-		
	18	Release production in ERP	3		X
	20	Fill out Q-document	7		
	new	Check tool	3		
	13	Get replacement hose from store	8		X
	14	Mount spare hose	6		X

Total time required in min.	Now	Before
Internal	62	124
External	80	15

Classify and Optimize Setup Steps

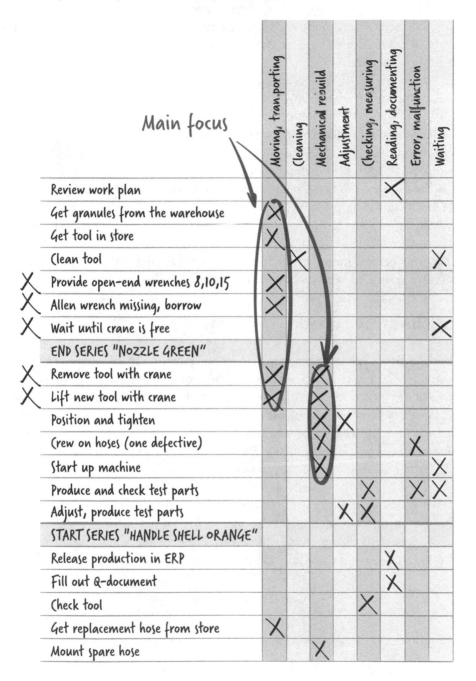

Main focus

	Moving, transporting	Cleaning	Mechanical rebuild	Adjustment	Checking, measuring	Reading, documenting	Error, malfunction	Waiting
Review work plan						X		
Get granules from the warehouse	X							
Get tool in store	X							
Clean tool		X						X
Provide open-end wrenches 8,10,15	X							
Allen wrench missing, borrow	X							
Wait until crane is free								X
END SERIES "NOZZLE GREEN"								
Remove tool with crane	X		X					
Lift new tool with crane	X		X					
Position and tighten			X	X				
Crew on hoses (one defective)			X				X	
Start up machine			X					X
Produce and check test parts					X		X	X
Adjust, produce test parts				X	X			
START SERIES "HANDLE SHELL ORANGE"								
Release production in ERP						X		
Fill out Q-document						X		
Check tool					X			
Get replacement hose from store	X							
Mount spare hose			X					

Now it is a matter of reducing the effort required for each individual work step. Here, too, you have discussed each production step in the team and classified them according to the various activities. Through this classification, the focal points jump right out at you. It is a decisive step on the way to finding a solution.

1. **Moving and transporting** are activities in which distances are covered, for example to fetch tools or devices.
2. **Cleaning of** tools, machines or even parts.
3. **Mechanically** rebuild. This is the core activity of the setup, for example, loosen the bolts of the tool.
4. **Adjust** the tool so that the machine is set correctly for the job.
5. **Checking** the parts for quality.
6. **Reading, documenting** quality parameters, processes or feedback in the IT system.
7. **Correct errors**: minor corrections or changes during the setup process.
8. **Waiting** until an activity has been completed.

Implement Optimization Measures

Now the times of the setup steps must be reduced. There are no limits to creativity when it comes to solutions. But you do not have to completely reinvent the wheel either. So, the team used some good solutions in the workshop, which were also incorporated into the LeanClean standards. While the organizational changes were easy to implement, the technical improvements to the individual steps cost quite a bit of money and time. A setup cart was organized so that tools did not have to be painstakingly gathered. Hoses were fitted with quick-release couplings. Quick clamps avoid screwing. And you have described every move in instructions so that no one takes an unnecessary detour during setup. At the next changeover, you are again ready with the stopwatch to record the effectiveness of the measures: 45 min, a new world record for the Plastmaster 2000!

Set-up trolley

Tools on site

Intelligent tools

Standards

Automatic tools

Short screw paths

Couplings
and quick clamps

Positioning aids

Slot instead of bore hole

Reduce connections

Form modules

Intelligent aids

Checklists always available

Self explanatory parts

Parallel setup

Summary of the SMED Method

The result of setup time optimization on the Plastmaster 2000 is impressive. Initially, the machine stood idle for 124 min until it was ready for the next part. After many improvements, the internal setup process now takes only 45 min. This was made possible by recording the actual situation and analyzing the individual steps of the setup process in the team. Organizational measures quickly showed their great effect. With further optimization tricks, the time could be reduced even further. 45 min is a great result for the first workshop but everybody is convinced that this can be improved even more. But the more optimized the process already is, the more difficult and intensive it is to save further minutes. Despite that we want to encourage you and your team to take on this challenge. The achieved target state is now the starting point for your next workshop. After all, the goal is SMED – single minute exchange of die! A vision you will never achieve but should guide you.

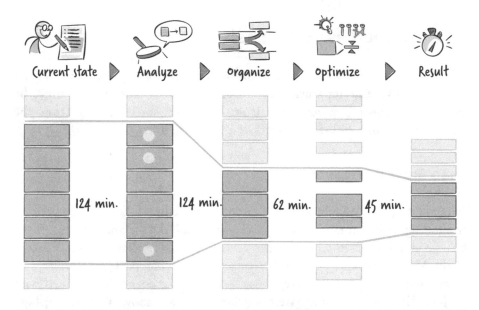

| Current state | ▷ | Analyze | ▷ | Organize | ▷ | Optimize | ▷ | Result |

124 min. 124 min. 62 min. 45 min.

Tip

1. Always start with a detailed recording of the steps of the setup process.
2. Check first whether activities can be performed when the machine is still running (external setup). Shift as many activities as possible from internal to external setup.
3. Look at the activities remaining in the internal setup and consider which measures can be used to improve them. The focus should be on improving the internal processes, because the capacity of the machine will raise immediately. The process for the external setup does not consume machine capacity but it is effort for the worker as well. So it must be reduced in the next step.
4. The machine operators are the changeover experts and have a broad technical knowledge that is indispensable in for optimization of the process. Therefore, they must always be involved in SMED. Listen to them carefully, and take them with you on the journey of setup time optimization.

5. Start with organizational measures to reduce internal setup time. When these measures have been exhausted, the next steps is thinking about technological changes.
6. Develop a standard template for the implementation of SMED. In particular, the time recording and the categorization should be standardized.
7. For more complex setup processes, it is helpful to record a video of the setup process. This way you can watch details several times and discuss them in the workshop.

4.5   Method 12: Lean Shelf

With smart shelves, you set the foundation for many Lean principles such as FIFO and separation of waste and value. The shelves must be designed accordingly. Removal from the front and loading from the back are only one of many aspects of a Lean shelf.

Lean Shelf Creates Order

Every time you walk through the production halls of LeanClean, you get the feeling that you are working in a gloomy warehouse and not in a modern production facility. Over the years, LeanClean has grown continuously. More racks and even higher racks were purchased, without a uniform standard and without a concept. Now even the last corner is crammed with shelves. They are high, they are bulky, and they are above all a sign of high inventories and inflexible production. They block the overview in production and the employees' view. These high racks are a symbol of separation and not of togetherness. This is no way to create teamwork.

First in – last out!
new material is placed
in front of the old

Collision between
logsitics
and production

After consumption, rear container
must be brought to front manually

On the way to a Lean company, these shelves will be an obstacle for you. Waste reduction will not work. The top levels can only be reached with a ladder. Since no clear storage locations have been defined, parts must be searched for again and again. Also, the FIFO principle cannot be kept because new material is always put in front of the old. Logistics fills the material in the shelve and production removes material from the shelf, and both use the same side. Collisions are pre-programmed. Can you see? you cannot implement the Lean principles this way.

There was an urgent need to develop a concept for a standard shelf at LeanClean. Together, you and your team set about building a prototype in the workshop. Each idea was questioned in terms of Lean principles. After several optimization loops, a concept was finalized that represented an ideal solution for everyone involved. And this is what the result looks like:

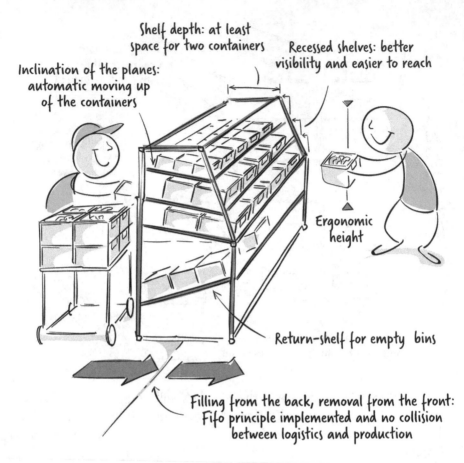

Shelf depth: at least space for two containers

Recessed shelves: better visibility and easier to reach

Inclination of the planes: automatic moving up of the containers

Ergonomic height

Return-shelf for empty bins

Filling from the back, removal from the front: Fifo principle implemented and no collision between logistics and production

You have included the new system in the "LeanClean Standards" and provided the new shelves for the first pilot area.

Are Shelves a Method?

Even though you will hardly find shelves as a method in Lean theory, they have a crucial function in Lean practice, especially when you manufacture lots of different products and variants. In our view, this qualifies Lean shelves as a top Lean method. In fact, all part delivery systems, most especially shelving, have a huge impact in implementing the 9 Lean principles. If you look at companies that have done a lot of work on Lean, you will notice that extensive effort has gone into the concepts of staging parts. Only with a well-thought-out shelf concept can Lean methods such as Kanban be implemented successfully.

> **Tip**
> 1. Unfortunately, there are no standard shelves from the catalog for all needs available. Therefore, get creative yourself. Use the internal competencies in development, maintenance, production or toolmaking to design the ideal solution for your production process.
> 2. First define standard containers for your production. The rack should be designed for these standard containers.
> 3. Build the Lean shelf from modular elements. There is a whole range of proven solutions for this on the market.
> 4. Create a prototype and discuss the result with all production and logistics employees involved. This gives you input that you would not get by designing a shelve at your office desk.
> 5. Go through the 9 principles and check how well your shelf meets them.

4.6   Method 13: The Milk Run = Takted Route Trains

The organization of the material flow is central to Lean Production. Timed routes ensure a reliable and predictable replenishment time for all customers of the logistic. Internally as well as externally.

The First Route at LeanClean Inc.

You observe the parts supply at the assembly area of the Elephant. There is a lot of hustle and bustle here. Forklifts hurriedly maneuver pallets in the narrow supply aisles and disappear back into the warehouse. Assembly workers regularly run out of parts and production comes to a standstill. The production employees then often go to the warehouse themselves and try to speed up the missing supplies. During this time, they cannot assemble, so it is a double waste.

You also notice the towers of empty packaging in assembly. These are first collected over the shift and then being returned in one go to the warehouse before closing time.

You can see that the parts supply at LeanClean is anything but efficient. There is a lot to be done here to eliminate the waste. Your task, together with your team, is therefore to bring flow and takt to the material supply. As always, you start your work with a thorough analysis to gain an overview of this logistics anthill. What are the transport routes? Where does the material come from and where is it consumed? How do we reduce distances in total?

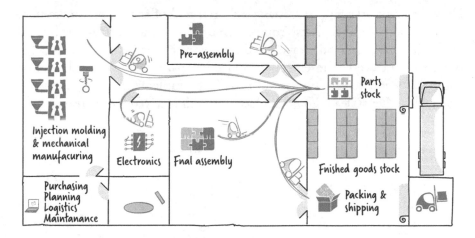

Many transports are carried out with forklifts, which drive the goods from the warehouse to the destination and usually returns empty to the warehouse. To visualize the current state, draw the observed routes on the floor plan of the warehouse. The star-shaped logistics from the warehouse leads to relatively long routes and poor utilization of the means of transport. Exact delivery dates, when something is delivered, do not exist here. Anyone who calls the warehouse and asks about the missing material gets the standard answer, "We'll bring it soon!". And soon means half an hour to three days. As a result, production also has no indication of when it will receive the parts delivery and can proceed with assembly. To alleviate the problem of arbitrary delivery times, LeanClean always schedules shipping dates to customers a little later and inventory a little higher.

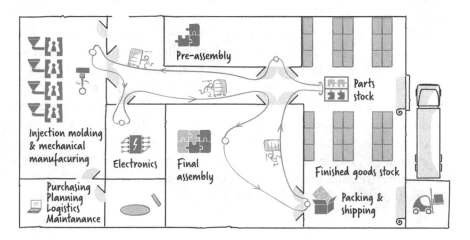

How can we keep transport distances shorter and offer the areas more transparency about delivery dates? You have your eureka moment on the bus ride home. The bus does not travel to the stops in a star pattern but connects them with a cleverly chosen route. And the departure times are also known and reliably adhered to. You would also like to implement such routes with a timetable at LeanClean. Your idea has been well received by the team.

Exchange empty milk bottle for a full...

In the new concept, you specify that logistics will initially supply the defined stops four times per shift. The picture shows the pre-assembly stop. Instead of using a forklift, now logistics brings the goods with a handcart. Each department has its own floor on this card. Handles arrive from the injection molding department with pinpoint accuracy, and the assembled handles are loaded and transported on to final assembly. Dizzyingly high stacks of empty containers are also a thing of the past. In the new system, when parts are brought, empty containers will be taken at the same time.

According to legend, the principle of "bring the full bottle, take the empty one back" was invented by a British milkman around 1860. In honor of the innovative milk logistician, the route train is therefore also known as the milk run. Even if the milk run is not your invention, it is a quantum leap for LeanClean logistics. You and your team have created a real cycle. Logistics now literally runs smoothly, in sync and without empty container jams. In addition, the production worker can now focus on his expertise, producing. And the logistician can use his expertise in transport logistics.

The Milkrun Conquers Supplier Logistics

As the origin of the Milkrun suggests, a clocked route can work not only in in-house logistics. Timed routes can also lead to significant efficiency gains and savings in procurement logistics (inbound logistics) or distribution logistics (outbound logistics). You remember the inventory analysis where we found that the critical parts come from four suppliers. In addition, three of them are virtually around the corner. Proximity is an ideal condition for an external Milkrun. Each plant still sends its own truck to deliver to LeanClean Inc. To ensure that the truck is always full, this only happens every three weeks. A perceived eternity for a company like LeanClean, which has the ambition to trim production to Lean.

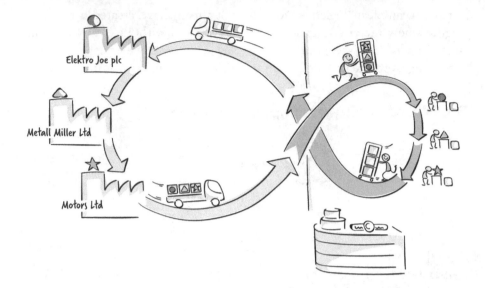

You and your team discuss logistics with the three suppliers right away. They are not happy with the three-week cycle either. After all, they are also accumulating material in their warehouses. Your idea of setting up a joint milk run for the three local suppliers is winning open doors. By bundling logistics, the delivery frequency can be tripled. Electrical components, metal goods and engines are now delivered weekly. Thanks to the closed-loop character, external logistics can also be switched to reusable containers. You no longer must dispose of cardboard packaging, and the environment is grateful too. The implementation of your idea shows that you have created a real win–win–win situation.

But that is just the beginning. With a daily delivery of the parts, even the intermediate storage at LeanClean could be saved. From the truck, it then goes directly to production. Just in time. So, the milk run not only saves you inventory and transport costs, but also intermediate steps and thus waste due to unnecessary processes. Just a few weeks later, the reusable containers are purchased, the processes defined, and the daily milk run is a reality.

Situation BEFORE With the Steps That Are Omitted

Flow of Goods With Takted Route Trains

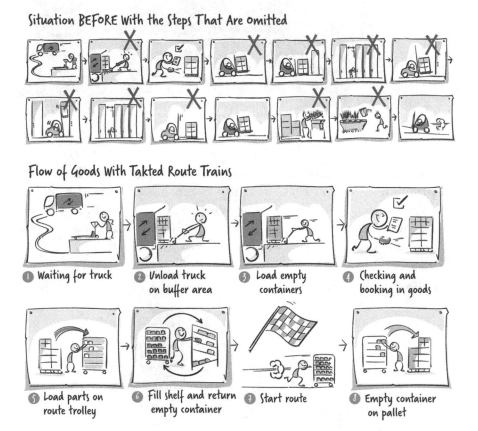

① Waiting for truck ② Unload truck on buffer area ③ Load empty containers ④ Checking and booking in goods

⑤ Load parts on route trolley ⑥ Fill shelf and return empty container ⑦ Start route ⑧ Empty container on pallet

You remember the handling level analysis you captured even before the Milkrun era? Now the question burns in your mind, did the Milkrun really bring an improvement? Once again, you set out with stopwatch and note-pad. After the recording, you are pleased and a little proud to see that only eight of the 14 handling steps remain. You were able to cut the effort almost in half. Put the beer on ice and invite everyone involved in the project to drink a toast. A success like this needs to be celebrated.

Tip

1. Get a picture of the material supply processes before implementing the routes. Draw the transport routes in the layout and record key figures that reflect the current state.
2. Start with a pilot project for internal delivery. Define the stops and draw the new route in the layout.
3. Define the frequencies in which the laps are run.
4. When the internal routes are stable, venture into the external Milkruns.

4.7   Method 14: Kanban

With Kanban you can organize the material supply according to the
pull principle. When a certain quantity is consumed, a signal to replenish is sent directly to the internal or external supplier. In this way, refill
is organized decentralized and quickly. The Kanban card is a common
means to transmit this signal with all necessary information.

From "Push" to "Pull"

The central production planning at LeanClean Inc. tries to provide the right
material at the right time for each workstation. For the injection molding
machine, the granulate; in pre-assembly, the handles, switches and screws;
and in final assembly, there are many other parts that have to be provided at
the right hour for each individual job. In theory, control works via a central
plan. In practice, this push principle approach fails big time at LeanClean.

The Plan   The Reality

Machine malfunctions, sickness absences, missing parts or rush orders throw the meticulously prepared plan into disarray. The result of this confusion of the plan by real events is higher inventories and yet postponed customer deadlines. But what exactly happens with push control? Let us look at how planning and control works today at LeanClean Inc. and what problems push control brings with it.

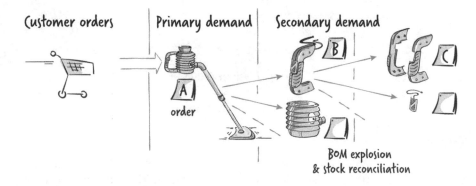

Planning begins with the determination of gross demand. This involves looking for the answer to the question of how many devices are likely to be sold in the coming period. This estimate is based on concrete customer orders, but also on an estimated sales forecast. From the gross demand, the still available stocks are deducted, resulting in the net demand, e.g., the number of devices to be produced.

First, it is about planning the equipment itself. Therefore, we are talking about primary demand. The net primary demand of the Elephants to be produced is 4000 pieces of orange, 2000 pieces of blue and 2000 pieces of green. To meet this demand, production orders must now be created, and their respective processing times calculated. Then this must be compared with the current capacity in production and the orders must be scheduled into the free time slots.

In order to determine the deadlines for the primary requirements, the individual parts of the Elephant, the so-called secondary requirements, must also be planned. This is because every handle assembly, every housing part and every electronic component of the Elephant must also be produced, pre-assembled and ready for final assembly on time. For this purpose, the bill of material is broken down into its individual parts. Assemblies such as the handle are in turn broken down into their component parts: left handle, right handle, switch, display and screws. When there is nothing left to disassemble, the gross dependent requirements planning is done. We determine the production quantity, the so-called net secondary requirement, considering the available stock. And finally, all these parts must be produced or purchased on the right date.

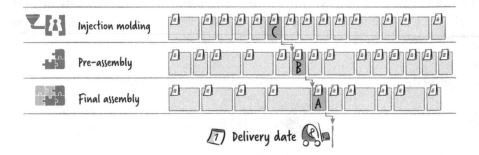

The Plan

Some departments, such as injection molding production, do not only produce parts for the Elephant. Scheduling and capacity planning must therefore also fit in with deadlines for orders for other products. So, customer demand for the Octopus and the Snake also plays a role in planning. All these requirements must be coordinated and harmonized with each other. Without the help of IT systems, this would be impossible to manage. Fortunately, LeanClean has a clever ERP system that masters this task. As a result of the algorithmic fiddling, hundreds of interlinked production orders are created. From the date of the handle to the delivery date promised to the customer, everything is precisely timed for the coming period. So you ask yourself what could possibly go wrong?

Welcome to Reality

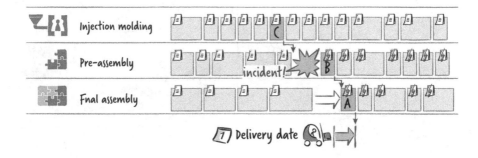

The plan is in place, but all the appointments are fatefully interdependent. Changing one deadline triggers a complicated chain reaction. At LeanClean, there are plenty of reasons for changing job deadlines: missing parts, sick leave, rush jobs, plant breakdowns or quality problems. This time, for example, it was a sick worker in the fully loaded pre-assembly department who threw the meticulously prepared plan out of kilter. This not only delays subsequent orders. The handle housings have already been produced and now cannot be installed according to plan. There is no space in the pre-assembly department; they cannot stay there. Therefore, the parts must be transported to the warehouse and stored there. This requires additional handling and puts additional strain on the warehouse, which is full to bursting.

On the planning side, all the due dates for all departments and all parts must be rescheduled in another calculation run. We very much hope that this time everything will go according to plan and that no case of illness or rush job will bring the house of cards crashing down. Yes, as we all know, hope dies last.

Replenishment for the Handles

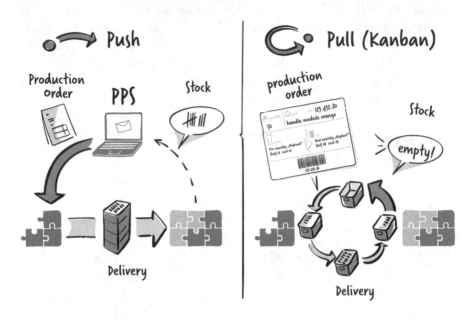

The push approach, where every order is created and scheduled centrally, has proved its vulnerability. What could be the alternative? Is there a system that generates the production orders itself, is oriented to the effective demand according to pull and can cope with the surprises of a production? This method is called Kanban.

Kanban organizes production control according to the pull principle and makes the logic for replenishment transparent and comprehensible for everyone. If you use Kanban correctly, you will reduce inventory, shorten lead times and regain control of supply security. Speaking of getting a grip: the handle assembly would be an ideal first Kanban project.

With Kanban, the production order is not started by a central planning. The idea of Kanban is that an empty container immediately triggers a production order itself. If a container with handles becomes empty in final assembly, it is sent to pre-assembly by the Milkrun. And in pre-assembly, the Kanban label on the container informs that it wants to be filled again with handle assemblies in orange. This wish is an order! Pre-assembly fills the container immediately and thus reacts directly to consumption. There are Kanban systems in which the cards are detached from the empty container and transported separately from it to the refilling location. In our system, the label is permanently attached to the container. The number of cards is therefore equal to the number of containers.

Two? Three? Or Ten Containers?

When a container runs low in final assembly, a full box must of course already be available as a replenishment. Thus, at least two containers circulate in a Kanban control loop. However, the number of containers must be calculated individually for each control loop. It depends on the consumption rate, the replenishment time and the number of parts per container.

To calculate the number of containers (according to Schönsleben 2016), this formula applies to our case (Kanban is dispensed according to consumption):

$$number\ of\ container\ C = \frac{1 + consumption\ rate\ V_{max} * replenishment\ time\ t_{max}}{parts\ per\ container\ n} * safety\ factor\ S$$

or abbreviated

$$C = 1 + \frac{V_{max} * t_{max}}{n} * S\ being\ C_{min} = 2$$

How many Kanban cards, respectively containers, are now required for the control cycle "handle assembly orange"?

V_{max} is the maximum consumption rate of handles. In two shifts, e.g., in 900 min, a maximum of 200 orange handles are consumed in the final assembly.

$V_{max\ orange} = 200$ handles / 900 min = 0.22 handles per minute.

t_{max} is the maximum replenishment time. This is where the Milkrun proves its worth. It delivers new handle assemblies punctually every two hours (=120 min), picks up the empty containers and later returns the full ones (=120 min). But the transport is only a part of the replenishment time. In addition, you must add the time needed for screwing the handles together in the pre-assembly department. The pre-assembly works in only one shift and the final assembly in two shifts. So, we must assume that the containers with the Kanban card are replenished every two shifts (= 900 min) in the pre-assembly. The maximum replenishment time is therefore 900 min + 120 min + 120 min = 1140 min.

n is the quantity of handles per container or per Kanban card. How big you choose the container depends on the space available in the assembly and the ergonomics. You and your team tried different containers and chose one that fits 50 handles. The full bin weighs just under 7 kg, so it is not too heavy. In the new standard shelves, all the bins for the handle variants have space in zone A. To ensure that the handles reach the end customer without scratches, the bins are equipped with padded inserts. So, you can solve the problem of scratching the handles, at the same time.

The last factor of the Kanban formula is **S, the safety factor.** Depending on how uncertain the replenishment time and consumption is, a safety factor should be considered. Since we have hardly any risks with the handles, you can plan with a safety factor of 10%. The more stable and smoothed the replenishment time is, the smaller your and t_{max} the less stock and cards you need in a cycle.

This then results in the following number of containers for the control loop of the orange handles:

$$C_{orange} = 1 + \frac{0{,}22\,Handles/min * 1140\,min}{50\,Handles/Container} * 1{,}1 = 7\,Containers.$$

With the same formula you can now calculate the number of containers for the handles in blue and green.

Here, the consumption is 0.11 handles per minutes each.

$$C_{blue} = 1 + \frac{0{,}11\,Handles/min * 1140\,min}{50\,Handles/Container} * 1{,}1 = 4\,Containers.$$

$$C_{green} = 1 + \frac{0{,}11\,Handles/min * 1140\,min}{50\,Handles/Container} * 1{,}1 = 4\,Containers.$$

The quantities and containers are now determined. Now you need to design and print the Kanban cards. The card must contain all the necessary information for replenishment: the article, the place of production and delivery, and the quantity per container. Scanning the barcode on the card makes it possible to book the produced quantity directly into the ERP.

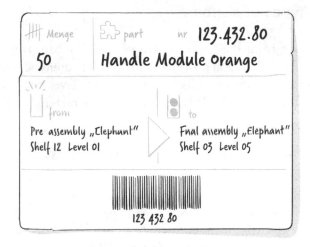

The pilot project for the handle was quickly implemented, and the benefits of the new system were clearly noticeable and measurable in production. One effect that was to be expected is the inventory reduction of the handles. If all the bins were full, the worst case for inventory, you would have 750 handles (50 handles/bins*15 bins) in the Kanban control loop between assembly and pre-assembly. At the time of value stream mapping, there were still 1629. Replenishment control now works decentrally and no longer depends on complex schedule control. Employees in final assembly no longer must complain about missing handles. Logistics processes have also become leaner, because thanks to direct delivery via Milkrun, the handles no longer must be stored and retrieved in the central warehouse. This saves countless handling steps, a heap of administrative work and valuable storage space.

Full Shelves – Like in the Supermarket

Like customers in a supermarket, the employees take the full boxes from a shelf. From their point of view, the shelves are always automatically filled, no matter what they take out. Just like it is the case with your supermarket around the corner. That is why this shelf, which is supplied with handle assemblies and other parts according to consumption, is also called *supermarket* in the Lean Production concept.

The Three Types of Kanban

As you could see from the handle example, the Kanban principle is quite simple in contrast to the central, administration-intensive push concept. If the stock level in a process falls below a certain level, the supply process receives a signal to replenish. This ensures that the material is always available for the customer.

This Kanban logic is very flexible to use. You can control different processes for the replenishment of parts in this way. We distinguish three Kanban types:

1. **Production Kanban**
   This is the type of Kanban we used to control the handle. The Kanban signal triggers a production order for replenishment.
2. **Transport Kanban**
   The part is available in a warehouse and the Kanban signal triggers a transport process for replenishment.
3. **Supplier Kanban**
   The Kanban signal triggers an ordering process with an external supplier for replenishment.

The Kanban Signal

The signal for postproduction can be generated fully automatically and digitally. For example, by a scale that measures the stock of a container in real time and sends the Kanban signal when the weight falls below a certain level. The signal can also be designed as a card on the container, as in our example, or it can be transmitted just verbally and on demand. Each type of signaling must be ideally matched to the process. All signal types have in common that falling below a certain stock level is the trigger for replenishment.

The Kanban signal in the form of a card is quite common in practice. It contains the information for replenishment and is physically transported from the place of consumption (sink) to the place of production (source). In a complex production with an exceptionally large number of parts, the card has the advantage of carrying all the necessary information about the item to be re-procured. Even with many thousands of parts, it is possible to control replenishment with Kanban cards without any problems. Often the question still arises in practice whether the card remains fixed to the container. There is no right or wrong. It depends on your situation. It is important that you think about in detail how the cards and containers are organized and how the process looks like.

By using barcodes or RFID on the Kanban card, you can digitize the bookings for the material flow in the ERP system and speed up the process. If the Kanban card is already scanned as empty at the point of consumption, the information reaches the subsequent delivery process in real time and not only when the empty container with the card arrives.

How should you get started with your Kanban process? With a simple card or with barcode and scanner? Usually, it is advisable to start with a card system to benefit from the advantages as soon as possible. In a second step you can always optimize the system with the possibilities of barcodes, scanners and all the digital wonder tools.

When Does Kanban Make Sense?

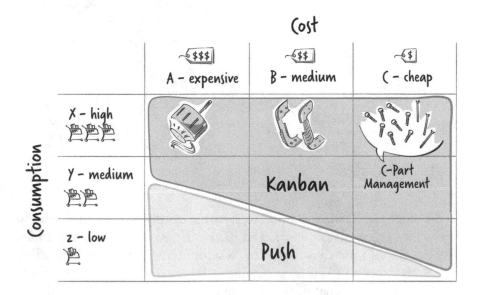

There are cases in which parts are used only rarely, in extreme cases even only for a single special order. It would not make sense to keep a container with such parts permanently in stock via a Kanban control loop. So, for which parts does Kanban control make sense and for which parts does it not? In the inventory analysis you have classified the parts according to the ABC–XYZ logic. You can now use this classification to derive a trend.

Basically, parts with a high and medium consumption, e.g., the X and Y parts, are predestined for a continuous control loop, such as Kanban. For A–Z, such as the gripper arm for the "Octopus", Kanban would be the wrong method. The low consumption and high cost would result in high values going unused. Demand-driven provision via push is usually better for these parts.

C-Parts Management

On the other side of the spectrum, we have inexpensive parts, but due to the variety, they cause a lot of effort. This parts such as screws, nuts or washers, cost little per unit but cause considerable administrative and logistical effort. To keep the stock of C-parts on reasonable levels, you could buy the screws in small quantities. Small quantities mean ordering frequently. Quickly, the administrative costs incurred for each order exceed the material value. Does the recipe now mean buying large quantities, preferably a year's supply? That would then cause our storage space to burst and is therefore not a Lean option. This is where Kanban comes into play in a special variant.

It is no wonder that the administratively Lean Kanban for small C-parts quickly established itself. An external partner ensures the supply. LeanClean obtains its small parts from the company called C-Part-Partners. They have set up a C-parts supermarket in each department. The workers fill up their containers for the assembly area here as needed. If a Kanban container is empty, the worker scans the box, which immediately triggers an order at C-Part-Partners. The latter leaves the tour at fixed intervals and brings a full crate of this type to the next tour and picks up the empty one. Due to his (very) wide range of products and his many customers, the C-Part supplier can bundle the quantities and utilize the routes well despite frequent deliveries. For you, this concept means that everything from the order to the provision on the shelf is handled from a single source.

Tip
1. Start with a Kanban pilot project and measure the outcome based on inventory levels and the effort required for transportation and picking.
2. Use a little more safety in the pilot project and equip the control loop with more containers than would be mathematically necessary. In this way, you take away the employees' fear that the material will run out. If the processes prove to be stable, the number of containers in the control loop can be reduced later.
3. All departments and employees involved, as well as management, must understand the Kanban method and be trained accordingly. Errors in the Kanban process can quickly bring the material flow and production to a standstill, for example if an employee returns an empty container with a long delay. And express orders from management can bring any Kanban control loop to its knees.
4. Define the layout, content and format of the Kanban card and set the standard for it. Provide the card with a QR code, barcode or RFID to automate the booking process. Most ERP systems support the Kanban methodology.
5. Define clear responsibilities for the Kanban process. A Kanban control loop only works if the consumption rate V_{max} is never higher than defined and the replenishment time is never t_{max} exceeded. These factors must be monitored. If they change because of changes in consumption or as a result of longer replenishment times, you must adjust the number of cards or the quantity per card.

4.8 Method 15: Line Balancing

With line balancing, you distribute the work content ideally and according to the takt principle to achieve a balanced and optimal Operator Balance Chart (OBC).

Balancing the Elephant Final Assembly

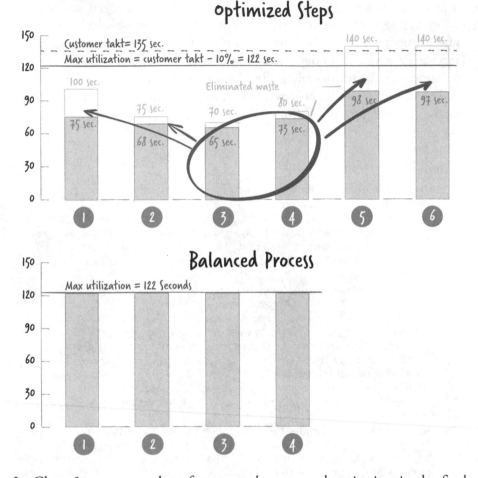

In Chap. 3, you saw a lot of waste and poor synchronization in the final assembly of the Elephant on the operator balance chart (OBC). You were able to eliminate the overloading of workstations 5 and 6 thanks to optimization of the paths with zoning and better organization through 5S. All steps are now working below the customer takt, which does not mean that the process is efficient. Even after optimization, the workstations are very unevenly utilized. The worker in step two still spends too much time waiting and overproducing instead of doing value-added work.

The challenge for you and your team now is to divide the workload more intelligently. That is exactly the goal of balancing. You distribute the work content in such a way that each workstation comes as close as possible to the customer's takt and can still always meet it. You take an iterative approach to balance the line. Each idea for a better work distribution is recorded in the OBC, discussed and then tested. If the test is successful, you will implement the idea in production. This is time-consuming, because with the new distribution of work steps, the material must also be provided differently and the tools adapted for each workstation. So, you shimmy forward with the team from work content to work content and improvement idea to improvement idea. In the end, you have a well-timed final assembly, which now only needs four workstations instead of six. With your project, you have saved a third of valuable resources that LeanClean can urgently use in other production lines in view of its rapid growth.

Tip

1. Balancing is always teamwork. The production staff but also the logistics and quality department should be represented here.
2. Make sure that a neutral moderator leads the workshops. Otherwise, you run the risk of your team getting tangled up in detailed discussions and not achieving their goals.
3. It is not possible to handle day-to-day business and participate in a line balancing workshop at the same time. Each team member who is involved must be released from all other duties during the workshop.
4. Start the first day of the workshop with training on waste and Lean principles so everyone is talking about the same thing.
5. Test ideas for synchronization as realistically as possible. Improvise equipment, shelves or trolleys that you need to implement the idea but do not yet have. Use cardboard, tape, wooden slats or whatever else you have on hand.
6. You cannot always complete the implementation within the workshop. Create an action plan to follow up on the open items.

4.9   Method 16: Set Building

With sets, you can intelligently bundle the supply of parts and keep inventory within limits in the area of added value. Instead of individual parts, an entire set is transported. This makes logistics more efficient and the walking distances for the worker shorter at the same time.

The Octopus Set

The Elephant is a relatively simple product with few parts compared to the Octopus. Containers with handles, motors or housings can be provided directly at the workstation. With the Octopus, however, with its five different functions, the numerous parts would take up a lot of floor space. Thus to gather the components for five arms, the worker would have to walk long distances.

Kits for final assembly

The idea behind the "set building" is that logistics put together a kit with the most expensive and largest parts (category A and B) for the Octopus. Small parts and screws are excluded here because they take up little space in assembly anyway. Logistics could then make these sets available to production via a Kanban control loop. In addition to reducing search times and routes, the risk of incorrect parts being installed can also be reduced.

One question remains that is always being asked in connection with sets isn't the waste simply rolled onto logistics now? Additionally the concern is openly debated if the logistician can just as easily put the wrong part in the set?

These concerns have some validity. But only partially. Although the logistics effort increases, the savings are still more significant. A logistics specialist can also prepare several sets at the same time, which means less searching and handling for individual parts. In addition, the layout in the supermarket can be optimized precisely for set formation, which would not be possible at the assembly workstation. Once again, set formation also realizes the principle of separating value creation from waste.

Tip

1. Use set building when space for material staging leads to walkways in the value-added area. Sets make sense, especially when many and large parts are in play.
2. Assemble a set of these parts and simulate the assembly process.
3. Ideally, if all the parts in the set come from *a single* supplier, the set then can already be assembled and delivered complete by that supplier.
4. If the parts come from different suppliers or are produced internally, design and test an arrangement in the supermarket that optimally supports the picking of the set.
5. Measure the time it takes to build a set in the logistics supermarket. Then compare these cost of set building with the savings in assembly.

4.10 Method 17: Sequencing

Sequencing is the sequence planning of orders. Sequencing allows you to provide the parts in a pre-defined and ideal order. This means that only the parts that are needed are kept in stock. Necessarily in the right quality. In this way, you achieve flow and avoid an artificial rush.

A Chain of lucky Orders

Through set-building, you and your team have implemented another streamlining of the production. But you can not rest long on your laurels. Consumers and the Octopus fan community are increasingly demanding more than just a device in standard configuration and in the standard orange color. The color palette is to be expanded to 10 colors. In addition, three more features are to be configurable via the online store. The eagerly awaited additional features will be presented to the press at the World Cleaning Show in Las Vegas. New to the range are the mop for shiny floors, the saucepan scrubber for cleaning charred pans and the violin robot with Stradivari qualities, which turns every candlelight dinner into an experience with romantic Vivaldi tunes.

Despite all the technical refinements and innovative functions: the price is also under pressure here and the demands for a fast delivery time are high. That is why you and your team are entrusted with the task of setting up a logistics concept for this variant-rich production.

A classic provision of the material in the assembly you have checked. But it quickly became clear that the version in 10 colors with three different additional features would lead to 30 different sets. It would be unthinkable to supply all these set variants in final assembly by Kanban. Too much space, too long distances and too much inventory would be the result – waste, which we so eagerly fight against.

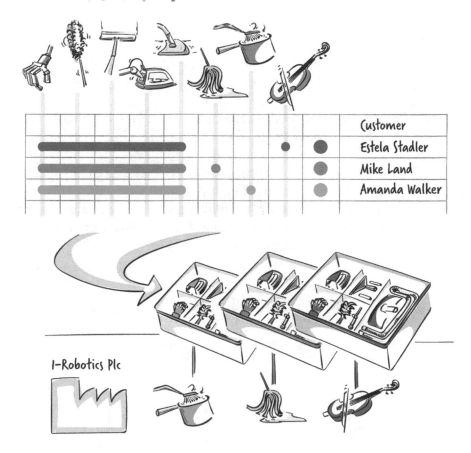

Your approach is therefore a combination of one-piece flow and set building. The information about the variant of each order goes to the Octopus supermarket. There, the set is equipped by logistics with the five standard features in the desired color. Now the question for you and your team is how you could provide the additional features, all three of which are supplied by Rolf Robotics: into the set or could the parts perhaps be delivered directly, in sequence, to final assembly?

Just-In-Sequence (JIS)

Instead of placing each part in the warehouse separately and sorted by type, the parts are to be delivered directly to production in the correct sequence as the customer orders are received. Since.

Rolf Robotics must first provide the correct feature, e.g., mop, saucepan scrubber or violin in the correct color, and another day passes before the delivery arrives at LeanClean, the planned sequence must be frozen two days in advance. This sequence may not be changed at short notice thereafter (not even by management) if the sequencing is to work.

The more stable your process is, the longer you can "freeze" the sequence. This gives your suppliers more lead time and enables remote plants with long transport routes to deliver according to the just-in-sequence principle. This opens great potential for you.

Do Not Step Out of Line!

Sequencing leads to the radical reduction of stocks and waste. But it also places high demands on process stability. Missing parts and quality problems can throw a spanner in the works of your planned sequence. If the sequence of supply to the sequence of assembly shifts by just one cycle, Mike Land`s configuration is mounted on Amanda Walker`s Octopus housing. A huge chaos, which would bring costs, overtime and delivery delays. This means you must ensure absolute process stability to enable production in sequence. Squeezing rush jobs in between is not only undesirable here, but strictly forbidden. Therefore, the sequencing method, even more than the other Lean methods, requires the understanding and commitment to change of all employees at all levels.

When Is Sequencing the Right Choice?

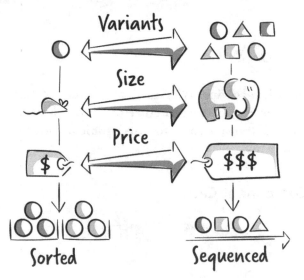

Sequencing is the consistent implemention of the 5R principle. In other words, the provision of the right part, in the right quality, at the right time, in the right quantity and at the right place. Many processes work in batches, and subsequently arranging the various part variants into a sequence involves a lot of effort. Sequencing pays off, especially for expensive and large parts with high consumption and many variants that are installed at the same workstation. The Octopus, with its many colors and three configuration options, is thus predestined for just-in-sequence provision.

Sequencing is possible in combination with set formation. Then the set can be picked in the right sequence for the right product. As our example shows, it is also possible to deliver individual parts, such as the additional feature, in sequence from a supplier.

Tip

1. Sequencing requires stable processes. If missing parts or a high error rate still occur regularly, sequencing becomes difficult.
2. Use inventory analysis to identify large parts with many variants and high inventory. These are potential candidates for sequencing.
3. Establish the planned sequence of production and measure its adherence over several weeks. Work on sequence stability until it reaches a sufficiently high level.
4. Calculate what costs could be saved by sequencing and what additional costs would be incurred.
5. Make the sequence internally by picking in a supermarket or already directly at the supplier.

4.11 Method 18: A3

Structured problem-solving is critical to success of Lean Production. The A3 method gives you this structure on an A3 sheet. In this way, you bring system and transparency to the problem-solving process in all areas of the company.

The Soft Component Case

The more you live the Lean philosophy in your company, the more problems but also ideas for improvement will surface. Some ideas, such as labeling or floor maching in a 5S workshop, you can implement immediately without much effort. But you will also encounter more complex problems whose root causes lie buried deep. Only when you bring them to light, you can find a solution. Here, a structured problem-solving process will help you.

Solve Problems Systematically

A simple sheet of paper with a given structure can save you from rash and unstructured actions. In fact, A3 is a paper format (29,7 cm × 42,0 cm) which gave the name to this method. Though the paper standard has its roots in Germany in the 1920s it is quite common in the rest of the world. And this format fits perfect for this method. This is how it may have evolved as a name for a classic Japanese Toyota Lean method. In this sheet of paper, seven methodological steps lead you from analysis to solution and its evaluation.

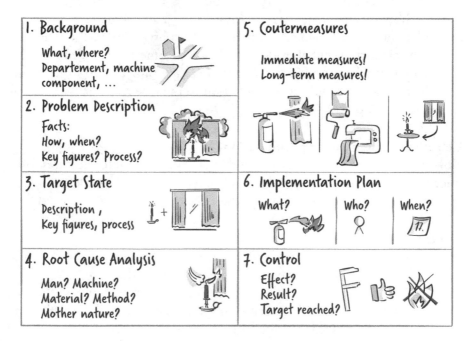

Background
Before we get into the problem, let us briefly define the context. Where does the problem occur, on which component and which machine? In this way, we ensure that everyone involved is mentally in the same place.

The Problem Description
The starting point of any problem-solving is an *accurate* description of the case. What happens when? This is about a precise description of the problem, backed up with figures, data and facts. A precise description of the actual process is also part of this.

Define Target State
While the team now agrees on what is going wrong, it also needs clarification on what the target state should be. This sounds trivial, but that does not mean it is unimportant.

Root Cause Analysis
Now the problem and the process are described so precisely that there are some candidates for suspicion. A few hot leads in the search for the culprit are human error, machine malfunctions, material defects, errors in method or subtle environmental influences. To substantiate your suspicions, you look for circumstantial evidence. Now it is time to start drilling: The root cause rarely lies on the surface, but is usually buried deep in the dark. Persistent questions about the why lead you to the true cause or even several causes for the disorder.

Countermeasures and Implementation
Now the causes are clear, and you can discuss suitable countermeasures. And since ideas are ineffective if they are not implemented, you also define who is to implement these measures and by when.

Control
An activity does not necessarily lead to the desired effect. To check whether your actions were able to solve the problem, evaluate the result thoroughly again. If it worked as desired, congratulations! And if there was no improvement, you go back to field four: cause analysis, countermeasures, implementation and checking again. May luck be with you this time.

Warrant for the Soft Component on the Handle

One of the first and most important problems you uncovered in the Pareto analysis is that the soft component on the handle housing can detach easily. You and your team want to get to the bottom of this problem, and the first thing you do is get a picture on site. An employee in the injection molding department is certain: "We've had a new low-cost supplier for the granulate for five months now. Since then, the problems have increased. Quality has its price."

What should you do now? Just quickly change the supplier?

To resist the temptation to jump to conclusions and fall into aimless actionism, you decide to solve the problem using the A3 method.

1. Background

Machine: Plastmaster 2000
Component: handle shell, soft component

2. Problem Description

Soft component on the handle
can be easily detached, causing 39% of scrap

3. Target State

Reduce scrap due to soft component
to below 20% by December

4. Root-cause Analysis

Granules: supplier has changed
Tool : is worn out
Machine : temperature fluctuates
Environment : humidity often above 65%

5. Countermeasures

Perform quality control of granules
regular maintainance of tool and
adjust maintenance schedule
Check temperature curve on Plastmaster
and readjust controller
Keep humidity at 50%-60%
Install dehumidifier

6. Implementation Plan

What?	Who?	When?
Granules	Mike	31. March
Maintenance Tool	Colin	31. March
Temperature Control	Stacy	7. April
Install Dehumidifier	Jim	30. April

7. Control

	Changes in Scrap
Granule	-0%
Maintenance Tool	-5%
Temperature control	-40%
Dehumidifier	-55%

In a workshop, you define the background, collect facts to describe the problem and agree on the target state that you want to reduce the scrap due to the soft component to below 20% by December. You discuss possible causes and determine four main suspects: the granules, the mold, the machine temperature or the air humidity are possible culprits. Thanks to the thorough preliminary work, the countermeasures are quickly taken and the persons responsible for implementation are noted down in black and white with a deadline.

Four weeks later you meet again and instead of assumptions you now have figures for your measures. The granule manufacturer is exonerated, and the mold is also only slightly to blame. The adjustment of the temperature control could solve half of the adhesion problem. The other half of the success can be attributed to the new dehumidifier system. So, you got rid of an annoying and expensive problem. Structured, efficient and always transparent for everyone involved. Respect!

1. Structure is central to solving problems. It is not enough to simply record the potential on a list. A responsible person for each measure must be defined, and the effect of the measures must be checked. This is the only way to successfully implement projects.
2. Use the A3 method to bring structure to problem-solving.
3. Make the A3 reports available to all stakeholders. This way, everyone has a good overview of the status of ongoing activities.
4. Training for the A3 method works wonders. But you do not need to plan a day-long training for this. In addition, you can methodically support the employees during the first application of A3.
5. Conserve your company's resources and do not start all initiatives and projects at once. Excessive actionism and too many open construction sites quickly lead to fatigue and frustration in the organization. Select a few important projects and lead them to success. This gives you energy again for new challenges.

4.12 Method 19: Poka-yoke

Poka-yoke assumes that users or people in general are not perfect and make mistakes. It aims to build in guardrails so that mistakes are noticed quickly or do not happen in the first place.

The Incorrectly Mounted Motor

The quality assurance department surveyed the assembly errors of the last month for the Elephant. With 23 cases, the error "motor cable incorrectly connected" made it to first place in the error statistics. Only in the final test did it become apparent that in each case the motor rotated in the wrong direction due to the incorrectly mounted plug. This mutated our vacuum cleaner into the ultimate leaf blower. While this may also be useful and could be an inspiration for the next product development, for our production it was just expensive, stressful and put the assembly out of sync due to all the reworks. You and your team are declaring war on these volatile errors.

Poka-Yoke for the Motor Plug

Translated from Japanese, poka-yoke means "avoid unfortunate mistakes" and is also a method from the Toyota production system. In the case of our misconnected engine cable, it is true that in-process measures such as checklists, warning signs or inspection instructions could help. However, the real goal of poka-yoke is that you develop a solution that prevents an error from occurring altogether.

Process

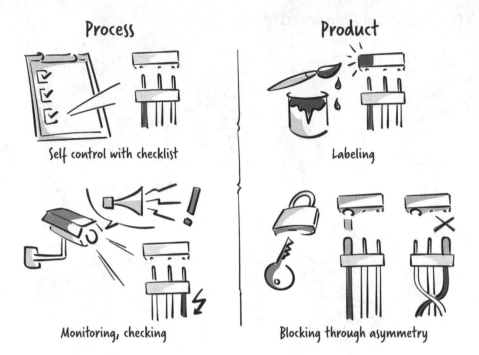

Self control with checklist

Product

Labeling

Monitoring, checking

Blocking through asymmetry

You could put a mark on the plug socket, where a wrongly mounted cable jumps to the eye. This helps to reduce the likelihood of an error, but not 100%. Therefore, the most effective way to mechanically prevent accidentally mix up is to use the asymmetrical shape of the connector. In our case, we chose to make the red pin of the connector slightly thicker. This way, it is simply impossible to insert the plug the wrong way around. This measure wakes even the most tired worker from his reverie on Monday morning and helps to ensure that only vacuum cleaners and not leaf blowers are produced in the future.

The most effective solutions usually require a design change of the product – and that is expensive again. Calculate the cost of failure by multiplying the frequency (23 cases per month) by the damage (42 $). This gives 966 $ damage costs per month. In this way, you can weigh up whether the measure is worthwhile to implement.

Poka-Yoke for a Carefree Life

Process   Product

Fortunately poka-yoke approaches are ubiquitous in everyday life. The ATM only releases the money when the bank card has been withdrawn. You are not allowed to board the plane until your ticket has been checked again. After thousands of misplaced and lost gas caps, this one is now hanging on a string: losing it is impossible. And how nice that a simple overflow in the sink ensures that water only flows into the drain and not into the whole apartment.

"A mistake like this must never happen again!" can be heard echoing through production halls here and there. Such announcements are not helpful. Accept that people make mistakes and put a stop to them with clever process or product design.

Tip

1. Rank your errors: how often does an error occur? What impact does it have?
2. First check the more effective variants of poka-yoke. Can the product be modified so that a form-fit solution prevents the error? Can an adjustment to the process eliminate the error? If neither is an option, can a warning label or checklist help prevent the error?

3. Supplement your standards with poka-yoke solutions that have proved successful in your business.
4. The poka-yoke mindset is important for the whole company. The close and
   early cooperation of development and production in the design of new
   products is the key to poka-yoke's success.

4.13 Method 20: Andon

**Andon shows you immediately whether all steps of your production are
in the green. Warning signals help you to identify problems at an early
stage, provide help quickly and rectify the fault in good time.**

Detect and Fix Faults in Real Time!

With the introduction of the takt, flow and pull principle, you have already ban-
ished a lot of waste from the processes and raised the variables of quality, delivery
capability and costs of the processes to a new level. However, a higher level also
requires greater stability of these variables. Process deviations caused by missing
parts, assembly errors or machine failures that upset the cycle no longer have any
place here. To detect disturbances in the process at an early stage, you need to
install an effective alarm system. The Lean method for this is called Andon.

Your first Andon project is starting in the injection molding department.
An Andon board is intended to make process deviations visible without

delay so that every employee can react to them immediately. In the injection molding department, you have therefore installed a large screen that is clearly visible from every workstation.

With your team you have defined three values to be displayed in real time on the Andon board. Firstly, a traffic light logic shows whether the Plastmaster 2000 is running (green), is stopped unplanned due to a malfunction (red) or is stopped according to plan (yellow). Secondly, a comparison of the actual and target number of pieces is displayed on the Andon board to monitor the adherence to delivery dates. And thirdly, it is indicated whether an employee needs help because he has a problem that he cannot solve on his own or within a reasonable time. In such a case, the worker can trigger this signal using an Andon button.

The Andon methodology helps to communicate the problems at an early stage and to approach the problem-solving process actively and quickly. For the team leaders, it is a means to get immediate feedback and to take a closer look at repetitive errors, for example with an A3 process, and to eliminate them sustainably.

After several weeks, a positive trend can also be seen in metrics such as OEE as employees, team leaders and department managers get a better sense of the issues.

Ultimately, the Andon method at LeanClean Inc. has also led to a more open approach to problems. Previously, reporting problems had the connotation of failure, blowing the whistle or denunciation. Through the Andon, uncovering and reporting process problems have been integrated into the production process and has become the norm. After one month in operation, the employees see that Andon is a real asset for LeanClean.

The Problem is Displayed – And Then?

You can show on an Andon display a digital, automatically generated signal (e.g., machine is running or stopped) but also one that is triggered by the employee to indicate a problem. In automotive production, you can see what is called an Andon pull cord on the assembly line in many factories. If the assembly worker pulls the chute, help is requested. Depending on the variant and company culture, pulling the Andon line results in an indication on the Andon board, the sounding of a signal or a complete stop of the assembly line. The idea behind it is simple. Do not accept mistakes, do not make mistakes, and above all, do not pass on a mistake.

The Andon should therefore first make it immediately transparent whether there is a problem or error in the process. But that is only the beginning. Rather, Andon is a control loop that only ends when the indicated problem has also been resolved. This means that the escalation paths and responsibilities must also be clarified in the event of a problem. Andon is thus a great help in implementing the 0-error principle.

By the way, an Andon is a traditional Japanese lantern with a wooden frame and paper window. Perhaps the first signal lamps that indicated the status of machines looked like these lanterns and thus coined the term. Meanwhile, Andon is known as one of the original methods of Lean Management.

Tip

1. Define the area in which the Andon method is to be applied. Consider which machines, which workstations should be included. Start with a pilot project and gather experience.
2. Consider which process deviations to look at and how to identify and display them in real time.
3. Displaying the problem, for example on an Andon board, is only the first step of the Andon process. Define and communicate who must react in case of a process deviation and how this intervention should proceed.
4. There are now good software solutions that can help you implement the Andon concept. They help you to retrieve the necessary data from various systems (such as ERP, machine controls or employee signals) in real time, process it and display it in the way you need for your Andon solution.

4.14 Method 21: KPIs

You may be familiar with the saying "what gets measured gets done". A fact-based approach is important to see and measure waste. For this purpose, metrics are suitable, with which you can directly or indirectly quantify the 7 types of waste and track their development over time. We are talking about so-called key performance indicators – KPIs for short.

Concrete KPIs for Each Production

While you use Andon to make the acute problems of your production transparent in real time, you can use KPIs to identify waste in the medium and long term. This helps you to track the effectiveness of the Lean methods already implemented. Furthermore, it is a basis for deciding in which area further Lean measures are necessary. KPIs are a gauge and guide for change. But which KPIs are suitable for your objectives?

Together with your team, you create an overview on a whiteboard of the KPIs from the categories of quality, time and cost that you think are particularly promising for measuring Lean progress at LeanClean.

Quality KPIs

Scrap incurs direct costs due to the disposal of defective parts, reproduction and the procurement of replacements. You should measure the cumulative cost in $. This is often better than in pieces. Money creates more concern. Rejects interfere with several Lean principles such as takt, flow or FIFO. In addition, your production times become unpredictable and methods such as sequencing become impossible due to a high reject rate

Customer complaints are poison for your reputation in the market and, like rejects, cause a great deal of administrative work. Directly or indirectly, they are also associated with high costs. The quality that the customer feels is always at the top of the hierarchy. This key figure is thus a measure of the quality culture and the implementation of the 0-defect principle. Only good equipment should leave the plant

By categorizing the errors, you get an overview of the frequency of each **type of error**. As we have demonstrated in the Pareto method, this is an effective step to fix the most common errors with priority and thus directly increase the quality

Yield is the percentage of good parts and the opposite of scrap. Yield is a good indicator of process stability and shows you the construction sites in your production

For the flow and takt principle, there is nothing worse than **missing parts.** These are parts that are needed but not available. Late delivery or loss due to scrap are just two of the many causes. Therefore, daily monitoring of these missing parts is important for the implementation of the Lean principles. Are you on the right track to eliminate these missing parts?

Flexible employees who are proficient in more than one machine or activity help you respond to fluctuations in production demand. You can quantify **employee flexibility** by analyzing how many cost centers an individual employee accounts for or how many workstations he or she controls on average

Time KPIs

The longer the **lead time for** a product or assembly to go through the production process, the higher the waste. Or put positively, the faster your parts go through a process, the leaner your production. Lead time is therefore an important waste indicator

How punctually are internal production orders completed? **On-time production** enables buffers and inventories to be kept low and customer orders to be delivered on schedule. If on-time delivery of production is low, the clients can only get products on time through intermediate storage. Since the start and end of the production order are usually always posted to the ERP system, the deviation from the planned date is relatively easy to measure

If the **takt time** is stable, for example in final assembly, the takt principle is also fulfilled. This KPI shows you the adherence to the takt time over a longer period

Many parts in LeanClean products, such as motors and switches, are sourced parts. In a Lean process without high inventories, a late delivery leads of these purchased parts often directly to delays in delivery of customer orders. To achieve more transparency about **supplier on-time delivery**, you measure the deviation from the planned date. Not only late delivery causes problems, but also early delivery disrupts the flow principle

On-time delivery of customer orders measures our own punctuality to the end customer. We demand punctuality from our suppliers, and logically our customers also rely on our delivery dates. But there are many reasons for delays. Errors, rework, waiting times, routes and transport mean uncertainties for on-time delivery. Hardly surprisingly, all of these factors are waste

The **order rate** shows you whether the customer orders regularly and smoothed, or irregularly and in large quantities. The reasons for the ordering behavior are not always driven by the market but are home-made. The order rate also gives you good indications of where you can still optimize the flow and pace

Cost KPIs

If waste decreases, costs also decrease. **Product costs** are therefore the ideal indicator for the progress of your Lean activities

Overall Equipment Effectiveness (OEE) measures the efficiency and effectiveness of a plant. Short setup and downtimes, as well as low scrap, lead to a high OEE. A consistently high OEE is therefore the key to high process stability and low process costs

The more firmly the flow, cycle and pull principles are anchored in production, the lower the inventory levels and the higher your **inventory turnover** will be. This key figure is therefore a good way of measuring progress in streamlining the supply chain

Productivity measures how high your output is in relation to your input, for example employee input. Higher output or lower effort leads to improved productivity and lower costs There is a direct correlation here: low waste leads to high productivity. Employee hours for a given product are a good measure of productivity. How many hours go into your product? What is the trend?

With **work-in-process**, we measure the stock used for production. With methods such as Milkrun, Kanban, SMED or sequencing, you can significantly reduce the goods in the process. With this key figure, you can evaluate the success in monetary terms

There is a lot of work behind the definition and later collection and evaluation of the KPI. It makes sense to select a few KPIs from the wealth of possible ones and to track them consistently over a longer period, e.g., months or years. You decided to start by measuring on-time delivery at LeanClean. You thought about this with your management team in a workshop and described the KPI in a comprehensible and clear way.

Definition of Delivery Reliability of Customer Orders

Name What should the key figure be called?	*On-time customer order*
Definition How is the key figure defined? With which variables and according to which formula is the key figure calculated?	*A delivery with a maximum deviation of one day from the originally promised shipping date is considered to be on schedule*
Reference Is it a relative, absolute or cumulative value?	*Relative, number of units delivered on time in % of total units delivered*
Data source How is the data collected? Which measuring points are used? Are the data already available or do they have to be collected additionally?	*From the ERP system, per order item committed delivery date and in shipping reported delivery date of this item*
Granularity and reference What is the granularity of the metric? Does the metric relate to a single part, an assembly, a product, a product line, a department or the entire company?	*Aggregated over each product line (Elephant, Snake and Octopus), The graph is displayed at the "flight level" product line, all detailed data for analysis is available and also supplied to the responsible employees*
Frequency How often does the key figure need to be collected? Daily, weekly or monthly?	*Monthly*
Representation How can the key figure be presented? Only as a number in a table or better visually in a diagram?	*Line graph, over the annual course*
Creator Can the key figure be generated automatically in an IT tool or must it be calculated manually?	*The key figure is automatically evaluated by the ERP system. The data quality depends on the clean feedback of the orders in the system. The graphical representation is done with a graphic software*
Communication How and where are the key figures communicated? Who has access to the evaluations?	*Online, available to every employee on the intranet*

Name What should the key figure be called?	*On-time customer order*
Destination Which target value is to be achieved by when?	*For each line 90% by December, every year this target value is discussed anew and adjusted if necessary*
Responsible Who is responsible for collecting, processing and communicating the KPI?	*Production Manager*

The many detailed questions were important, but more time-consuming than expected. The inclusion of the management team in the definition was also important to increase the acceptance of this important key figure.

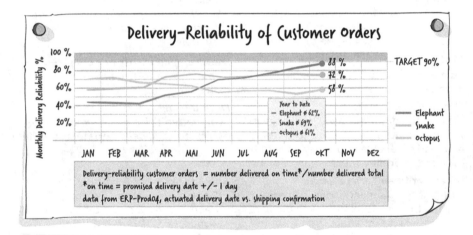

The collection of this first key figure involved a greater than expected initial effort. After all, the data in the ERP still had to be cleaned up so that it was suitable for calculating on-time delivery, a digital 5S action, so to speak. But then the feedback of the data in the ERP for the KPI succeeded without problems and with good quality and the KPI is eagerly awaited at the end of each month.

Transparency about the punctuality of deliveries to your customers has stimulated every department to reflect and optimize. In injection molding, scrap has been further reduced, final assembly has realized further Poka-Yoke-style improvements, and the shipping department has also streamlined its processes.

Thus, the entire company can now look back on a significant improvement in the KPI "on-time customer orders". The Elephant is already on the home stretch and the Snake and the Octopus are well on their way. How motivating and effective a simple and openly communicated percentage can be!

Importance of KPIs in Lean Management

In the Lean approach, the basic principle is that every change must have a *measurable and quantifiable* benefit for the process. KPIs are an indispensable tool for this measurement. They support you in uncovering waste selectively or in the entire process. Like a compass, KPIs also help you to target the right goals in the continuous improvement process.

Key figures create transparency and show you process deviations and problems. Instead of opinions, beliefs or gut feelings, fact-based KPIs now influence decisions at LeanClean. The prerequisite, of course, is that the KPIs are defined precisely and transparently and collected correctly.

At times, things can get a bit emotional in the company when your own beliefs are questioned by objectively collected KPIs. But at LeanClean, you have agreed to always approach changes in a fact-based, comprehensible manner and based on KPIs. This culture helps to introduce the KPIs.

For the KPIs to be accepted as a guide and target, there must be no debate about their correctness and no room for interpretation in the logic of the KPI. Unfortunately, this is complicated by the fact that there is no detailed, standardized and written definition for most KPIs that you can use universally. Rather, you must find, define and communicate the exact characteristics of the KPIs for your company and your processes.

Tip

1. Start with few KPIs. It is better to start with two to three key figures in order to keep the complexity small and to have the first KPIs available as quickly as possible.
2. Define the key figure clearly and precisely, avoid scope for interpretation.
3. Only if everyone can understand how the metric is arrived at will it be accepted. Always show the definition and structure of the metric with each publication, as we did in the footnote of the "On-time customer orders" example above.
4. Actively communicate the reasons why a metric is being measured. This creates clarity and trust. We do not want to use KPIs to control employees, but to improve processes together.
5. Presenting KPIs as abstract numbers in table form is often not enough. Visualize the KPIs in a simple diagram on which changes, and improvements are clear at first glance.
6. Use few and standardized color schemes, e.g., green for positive and red for negative deviations. But use only few colors, because too many different colors confuse more than help.
7. Make sure that the key figures are always up to date and widely communicated. Otherwise, they will quickly lose their importance and meaning.

8. Regularly review the key figures and their definition. Are they still appropriate for the change process?
9. Set goals for the KPIs. But the target does not have to be set right at the beginning of the measurement. Wait for a stabilization phase so that acceptance for the KPI can develop first. A periodic review of the targets is important.

4.15 Method 22: Shopfloor Management

Shopfloor management ensures that production runs smoothly and that Lean methods also function as they should in day-to-day operations. In short daily meetings, the status of production is discussed and, if necessary, measures are defined. By the way, this works not only in production, but everywhere in the company.

The Control of Lean Production

During your Lean projects, such as route optimization through zoning, logistics improvement through a Milkrun or the introduction of Kanban control loops for replenishment, you could count on a high level of commitment to the Lean theme. The team met and worked on the implementation of the method. This automatically provided the necessary focus and attention to make the planned change a success. After the happy conclusion of the Lean projects, you now must maintain the achieved level, even if Lean is now no longer such a prominent focus. To do this, it is necessary to regularly inspect the production processes and, if necessary, correct deviations. The shopfloor management method ensures that you stay in regular contact with the team and discuss the current issues.

The continuous maintenance program for the processes in LeanClean now takes place daily in the shopfloor management. To this end, you set up a 15-min meeting at 9 a.m. at which all decision-makers go over the most important events and the KPIs of the day. Here, the problems encountered in the Andon are discussed and actions on process deviations are determined. The meeting takes place in a strategic place, directly in the production area. 15 min is not a lot of time, but if the discussion is based on numbers, data and facts, it is more than enough. For this purpose, you and your team have set up an information point where all the necessary information is available to conduct the shopfloor meeting efficiently.

An up-to-date and transparent fact base is crucial for shopfloor management. In the first concept, the focus was deliberately placed on a few aspects.
Specifically, these are OEE, delivery reliability and yield. The 5S audits and
A3 reports are also posted here. Is everything in the green zone? Traffic lights
and visual management allow deviations to be identified quickly. The team
leader moderates the shopfloor meeting. For more complex problems, he
initiates a A3 process.

The benefits of shopfloor management are not complete and all-encom-
passing data or precisely informed employees, but the measures that are
jointly decided every day to counteract the current process deviations. After
all, it is only action that removes waste at LeanClean. A list of the measures
decided upon with their current status should therefore always be available
at the shopfloor meeting.

Tip

1. Set a fixed time frame for shopfloor management. Daily 15 min should be
   sufficient.
2. Evaluate which process deviations need to be monitored in the shopfloor
   management and which metrics should be used to do so.
3. Define a team for the shopfloor management. The participants must be able
   to influence the key figures.
4. What is important on the shopfloor management is that the team defines
   measures and tracks their implementation.

5. In the age of digitalization, information for the shopfloor can be created digitally and displayed on a large screen in real time. So that you do not have to update the key figures in paper form every day, you can think about such a digital solution. Some effort is required, but this digital process massively improves communication.

6. A temporal cascading over the different areas has proved to be target-oriented. Team shopfloor, followed by departmental shopfloor and finished with a daily shopfloor for the whole company, brings complete and fast transparency for every decision-making level.

5

Lean Change

Contents

They always say time changes things, but you actually have to change them yourself.
 Andy Warhol

5.1   Change Never Ends

How can you create a climate of change in your company to implement Lean successfully? What do you have to consider so that Lean also flies in your company? In the following chapter, we explore this "Lean change" question. We want to show you how to transform the readiness for change from "yet another new program" to "let's tackle this and succeed".

© Springer-Verlag GmbH Germany, part of Springer Nature 2022
R. Hänggi et al., *LEAN Production – Easy and Comprehensive*,
https://doi.org/10.1007/978-3-662-64527-7_5

Lean means a culture of constant change in small steps. This sum of many small improvements leads to a major change and success. Once you have implemented one change, plan the next change, because there is always waste that can be eliminated.

The basic idea that change must happen continuously is of central importance in the Lean concept. This is associated with the Kaizen management approach (in Japanese: continuous improvement) (Imai 1986). The continuous improvement process (CIP) in the Lean approach is therefore more than just a company suggestion scheme.

Change also never ends in Lean management because the perfect, completely waste-free process remains unattainable and always a vision.

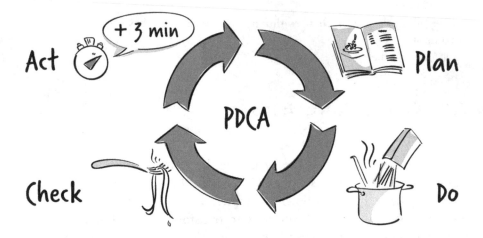

This way of thinking is also known in the PDCA approach from quality management. PDCA stands for "plan"–"do"–"check"–"act". The PDCA control loop says that you must first develop a concrete plan to solve the problem (plan phase). This plan must be detailed, implementable and understandable. The "do" phase is about implementing and realizing. The decisive phase is the next step (check). Check and measure whether the change has fulfilled the goal. In the last phase (act), the identified deviation from the target state from the "check" phase must be addressed.

Improvements should never be planned down to the last detail in Lean change. That would be paralyzing and costly. Instead, they must be constantly developed in a control loop. Implementation must thus happen in steps. The motto must be "better 60% now than 100% never". This brings fast results. But it also means that effectiveness must be monitored and readjusted after implementation. Thus, Lean cannot simply be implemented with one project. Lean is the result of a permanently running control loop.

5.2 The Change Control Loop

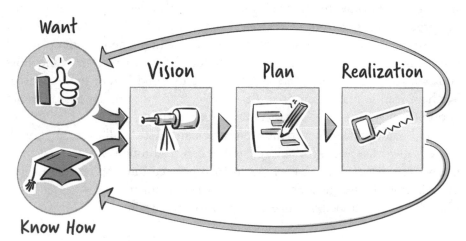

Lean is certainly not rocket science. Individual methods can also be applied simply and effectively on a selective basis. But a comprehensive change of processes in the sense of Lean in the entire company or even in a whole group is anything but simple. The change is a long-term task and an immense challenge. There are many stumbling blocks along the way that can make implementation difficult, and it is not uncommon for Lean approaches to fail because of these hurdles after initial successes.

Change can take a lot of energy, especially in the introductory phase. Transformation, the new and the unknown can be met with skepticism by employees. In addition, the term "Lean" is not new, and so preconceived opinions and perhaps even negative experiences from failed Lean implementations can slow down the change.

In retrospect of many good and some not so good Lean projects from all industries and cultures, we asked ourselves what was responsible for the success (or failure), because with seemingly similar basic conditions in the production process and product portfolio, we could see serious differences in the results.

The good news is that you do not have to leave everything to chance. There are many variables that you can actively influence to make the change a success. We have illustrated the dependencies in the change control loop.

The starting point of any change is to want it. This sounds trivial at first, but it is already the first major danger that change will not get off the ground. The motto in people's minds must be "we have to do something". In addition, knowledge about Lean is also necessary, because only those who want and know how to change something can also develop an idea of how the process works without waste (see also Liker and Ogden 2011). Now you can formulate a concrete vision of the future and implement it in a planned way. You need to say in this plan where and how you will start and how you will roll it out step by step. A plan is only good if it is implemented. We have seen so many times that Lean projects are not implemented despite good planning. The will to change was not great enough.

And because change never ends, the whole thing starts all over again, with more knowledge and with greater will, because through the change, new knowledge is acquired, which is again incorporated into the standards. The organization has learned. The new state is better but not yet good enough. The will to do even better triggers the control loop again.

5.3 Step 1: Wanting the Change

A Question of Attitude

Without wanting the change, the change does not pick up speed. Unfortunately, especially at the beginning of a Lean change, not everyone wants to change themselves and their working environment.

Lean management means changing all processes and thus the daily work of many employees, in virtually every area of the company. And if you go about it the right way, this change will not be temporary. So, it is understandable that it will trigger different reactions among employees. And so, you may find that the will to change is not equally strong in every employee. At the beginning of a change project, the will to change has four basic characteristics:

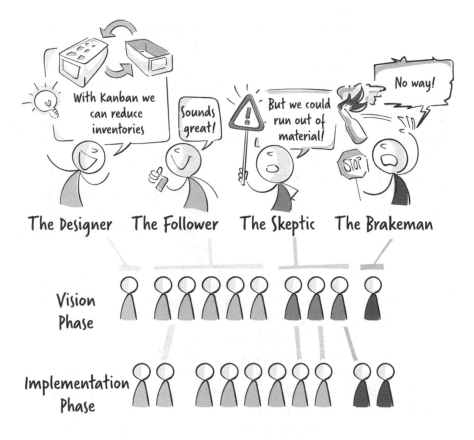

1. **The Lean designer** has a proactive attitude toward Lean change. He knows the Lean principles and methods and so the will to change is naturally great in the designer. He wants to project and tackle things himself. And he already has one or two ideas on how to redesign processes. You belong to this group.

2. **The Lean follower** has a basic knowledge of Lean and understands that waste is detrimental to the company. He therefore sees predominantly advantages in the planned Lean change and has a positive attitude toward

the change. However, followers do not initiate any improvements of their own accord.

3. **The Lean skeptic** has a rather negative attitude toward change. This can have two reasons. One is the familiarity and knowledge of existing processes and the fear of losing them after the change. The other reason could be that preconceptions about Lean have formed, stemming from a failed initiative or rudimentary knowledge. The skeptic tries to stay out of the change activities.

4. **The Lean brakeman** actively interferes with Lean implementation and uses his function to slow down innovation and improvement. If it is within the brakeman's decision-making range, he will talk down, defer or postpone improvement measures, because quick measurable changes from the Lean corner could make his previous work look pale. We can't look inside their heads, but we can only assume that the brakeman fears for his position of power or his reputation. Statements such as "Lean does not work for us because we are not a car factory", "Lean only works for products produced in large quantities", or "we are only a small fish for our suppliers" expose the Lean brakeman. During the implementation of Lean, new Lean brakemen may well develop. After all, there are processes that are eliminated and have a direct impact on the work of employees. This is often not appreciated and can lead to frustration.

A smart production manager, with lots of experience in Lean change, put it succinctly. "Lean is like any big change in operations. The 20–60–20 rule applies. 20% actively pull and change. 60% join in and trot along after the designer, and 20% just put the brakes on."

Of course, this categorization is very simplified and exaggerated, but it may help you find the way to bring as many employees as possible along.

We have seen many times that an initial Lean skeptic can turn into a Lean creator. Education, training and open discussions about Lean all play a role. In this way, employees learn to see the waste in their own processes and recognize the need for change. Over time, additional positive personal experiences can increase the will to change.

But for all the efforts to transport the will to change into the team, the starting point of the change is, however, in the management.

... First the Management Must Want!

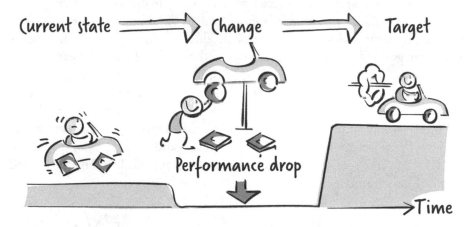

The will to change in top management is particularly important – even essential. The leadership must not only have the will, but also show it and communicate it again and again. Through words and through deeds. This does not mean that all ideas for change must come from management. Quite the opposite. But management must clear the way so that the entire organization can follow.

As the level of waste is lowered, new problems are flushed to the surface in the change process. It may even be that the performance of the process deteriorates until these problem stones are cleared out of the way. Especially in these times, management must not allow any doubts to arise about the will to change. Otherwise, with insufficient support, the Lean change will be aborted at the slightest imponderable.

That would send the wrong signal and would also be the wrong direction. It is part of the Lean culture that problems are made visible and quickly eliminated, because that's exactly how you become a Lean company. You fix the problems and generate flow of products or information. The bottle-necks are gone. These situations show whether the will for the Lean change is anchored.

The will to change is also strengthened by seeing waste in one's own pro-cesses. Regular on-site observations according to the motto "go and see" should therefore be part of the company culture. Does top management also have to be so deeply involved in the processes and their details? Yes, abso-lutely! Only those who understand the real process can lead effectively and continuously improve the process.

You will need time and resources in a Lean transformation. Here, too, the will to change in management plays a central role. After all, the resources must be made available and care must be taken to ensure that the priority for change is not lost in everyday life. For example, at a Swiss bathtub man-ufacturer, production stops for one day every two weeks. Every employee in all departments works together on the Lean transformation of the company on this day. By the way, even the boss pitches in on this 2-weekly Lean day.

The "love of problem solving" must be exemplified. Management should coach employees along the way and ensure that work is done as a team to implement. The will to change is thus exemplified to the team. Success will come quickly and then the change will be unstoppable. The train is running.

Create Your Personal Crisis with Facts, Figures and Data

The will to change is like the fuel for change – but how does it come about?

We think that the realization of the problems in the existing processes causes resentment. This is where the driving force for change comes from. This discontent can be broadly caused by a real crisis, for example when a company gets into financial difficulties. But you can also create the effect of the crisis artificially by making waste visible and communicating these grievances. You should therefore not wait until a crisis hits your company. Use the Lean methods to see the waste and the Lean KPIs for this purpose.

To cut to the chase. If your passenger tells you: "You are driving too fast", you will consider braking. If, on the other hand, he says: "You are driving 80 mph and only 30 mph are allowed here", your will to brake is certainly greater. Numbers, data and facts help you to objectify the problem and to strengthen the will to change.

5.4   Step 2: Build Knowledge

Train Every Employee

The will to change must be there. But this alone does not lead to Lean change in the company. Just as necessary as the will to change is the knowledge of Lean, its principles and methods. The will and the knowledge of how to do it are therefore the igniting mixture for change.

It is therefore important that you bring this knowledge into the company and anchor it. Comprehensive training on Lean basics, on waste and Lean principles is a first step in bringing all employees up to the same level of knowledge. Specific knowledge about Lean methods is also important and requires its own training. These should always be adapted to your products, processes and history of your company. You will build up a lot of knowledge in your own company from the change experience and this will in turn help to find new solutions.

Much general Lean knowledge is well documented, written down and accessible online. But perhaps most of the world's Lean knowledge is in countless smart solutions at thousands of companies that have been managing Lean for years. Some are proud of it and open their doors to visitors. Look around. Benchmark visits to companies are a good source of inspiration and a way to build your own practical knowledge.

Preserving the Knowledge in Standards

For an effective change, it is important to cement the Lean knowledge that has been built up in the form of standards. We have already learned about this in the standardization principle and discussed it in the 5S method. And because standards only make sense if everyone knows them, make sure that you communicate them.

Continuous improvement is the key mindset of Lean transformation. To prevent regression to the old processes, you need to capture the new knowledge in standards. These can be process descriptions, quality criteria, metrics, checklists, audits or even trainings.

It takes a lot of time and energy to work things out the first time. In particular, details that must be worked out at the beginning of the change, like the format of Kanban cards or the concept for shelves, are time-consuming. But so that you (and others) have this effort only once, you should set these "best practices" as a standard for your company, preferably even in printed form, for example as a small booklet. This way you develop your company-specific production concept. Standards not only secure the change for you, but also help you to roll out the change to all departments and locations of the company.

However, it is also important to find a healthy level of regulation when setting standards. You can also overdo it when setting standards and some flexibility and innovation must always be possible. At some point, ensuring standards cost more than they bring. Above all, standards must not be so narrow that you then need an exemption for every case. But on the other hand, they must be formulated and described so precisely that they can be implemented. The high art is then to formulate and present this description as simply and comprehensibly as possible.

And finally, a tip from the field. Involve other departments such as Marketing, IT, Finance or Human Resources in the creation of the standards so that the new process standards also fit their goals.

5.5   Step 3: Develop a Vision

How Sharp Is the Image?

If the employees in your company want to become better and know how they can achieve this with Lean methods, the change process can pick up speed. Now it is a matter of giving the change the right direction and setting guardrails. The idea of what the changed process must look like is the guideline. And if everyone has the same picture of the future in front of their visual eyes, they will also pull in the same direction.

Perhaps you have already had to endure management workshops in which generic company goals were worked out after many hours of work. "We want to be the market leader", "we want to leverage synergies at all levels", or "we want to maximize the shareholder value of our company" were hard-won outcomes. While these are all legitimate goals, they are not precise enough for a Lean change in manufacturing or the entire supply chain. That is because there would be an infinite number of ways to achieve these goals. In Lean change, it is more important to agree on the path than to describe a distant, abstract goal. To do this, the goals must become more concrete.

You therefore need a vision of the future that is primarily related to the processes in your company. Therefore, it is better to formulate milestones that are based on concrete operational goals within the process and its value stream and that focus on customer needs. For example, the vision should clarify the following questions: What delivery times are you aiming for? How many handling steps should there be in the future until the material is ready? What should the material and information flow look like? What lead time should the product have and what is the maximum amount of material that can be stored in the warehouse? How fast must the setup time be?

The vision must be communicated throughout the company and defined clearly and in detail for this purpose. Numbers can help to concretize the vision, but it is even better to transport the vision of the future in pictures, for example, as a value stream or swim lane diagram of the future, or as an ideal handling stage representation. What will the layout look like in the next product generation? The motto "a picture is weath a 1000 words" applies.

The Vision Does Not End at the Factory Gate

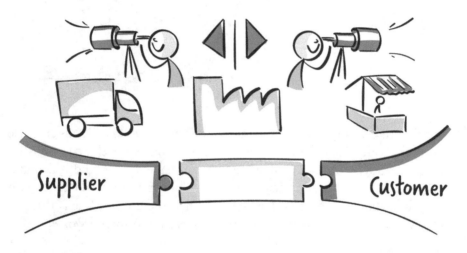

The production process is always connected to suppliers and partners. After implementing Lean methods such as Milkrun or just-in-sequence, suppliers are even more closely intertwined with your production process. Poor delivery performance then has a very quick and immediate impact on your own delivery performance. Therefore, early involvement of external partners is critical to success. Their input, as well as the need for their willingness to change, is important prerequisites for change. Experience shows that communication with partners improves through the joint development of a vision, and new win–win perspectives open up.

How Digital Is the Vision of the Future?

How do you combine Lean and digitization? This important but also conflict-ridden topic is playing an increasingly important role in practice. Although there are more and more possibilities to implement the production process faster, safer and with fewer personnel with the help of digitalization, what is technically feasible is not always Lean, sustainable and certainly not cost-optimal.

Especially when companies already have comprehensive IT systems in place to control production, there is a risk that a vision is developed about an information flow without considering the consequences on the real material flow. The result is many handling steps, longer throughput times and ultimately higher overall costs. For the customer, the bottom line is that the products become more expensive and the delivery performance worse.

Digital solutions hold enormous potential, but when it comes to creating the Lean vision of the future, the rule is: Lean first, digital second. This is not only because the result of this logic is a Lean process. This order also makes more sense for digitization, no matter what type and at what point in the production process. Digitization is always easier to implement with stable processes. Lean creates this stability.

Four examples of the right order in the vision:

1. After the changeover from push to pull, the Kanban control loops can be mapped digitally in the ERP. Bookings can thus be made by scan and in real time. And from the system, the history of Kanban bookings again provides important information for further optimization.
2. Sequencing brings great benefits but means effort for picking in the supermarket. The digital solution through pick-by-light systems directs the picker to the right bin and speeds up picking.
3. Shopfloor management brings the transparency needed to control production. Through a digital solution, KPIs can be made available in real time and without manual effort.
4. The broad collection of production and quality data in the Lean processes provides you with transparency and facts for further optimization. By storing and evaluating these data, you can identify further waste and derive improvements from it.

We can summarize that digitalization is not in competition with Lean. Digitalization is Lean's friend. So, integrate digitalization into the vision of Lean processes.

5.6   Step 4: Plan Implementation

Planning Is Teamwork

Once the goal and the shared vision are clear, measures must be developed and coordinated to turn this vision into reality. The implementation plan shows what will be implemented, when and by whom. The more employees are involved in the planning process, the higher the acceptance and the easier the implementation.

More extensive changes in processes that affect all departments should be planned for the long term and approved by the management team. Consider the composition of a steering committee that can approve the change plan but also call for it later. In any Lean transformation, it must be clear how the decision for the plan will be made.

But not every small change should be organized centrally. That would slow down the dynamics of change. After all, it is also the sum of many small improvements that make up the change. Each team should therefore have the competencies and the budget to implement Lean measures in its own area. The rules of the game, such as budget and capacity for this, should be clearly defined. Then small improvements can be implemented quickly, unbureaucratically and easily.

The plan and the measures must be communicated. This creates transparency and planning security and helps you to implement them quickly.

Workshops and Projects, but Also Quick Wins

For changes to be implemented successfully, an interplay of many factors is necessary. People, technology and organization must find each other anew, and familiar processes and procedures must be replaced by new approaches. For this to be realized, implementation in clearly defined steps has proved to be promising. Faster results, focused use of resources and immediate correction of errors are just some of the advantages of this approach. There are two organizational forms for implementing the change plan in steps: the project and the workshop.

In a project, a large, complex and interrelated task, such as a new plant or product, is organized. The project manager controls the schedule and resources of the project. Typically, it then runs for several weeks, months or years. For control purposes, the project is divided into phases, each of which concludes with milestones.

In a workshop, the task is limited and can be completed within one to two weeks. Many of the Lean methods presented can be worked on well in the form of a workshop. The workshop is "facilitated" by the moderator. The idea is that a problem is completed from analysis to implementation in one go. It is important for the efficiency of the workshop that you define the content, the process, the participants and the time exactly in an agenda, at least on an hourly basis. At the end of the workshop, the participants (not the moderator) present the result.

Applying a sequence of workshops per topic area, you can also implement large changes. Generally faster and more adaptable than through a project planning approach. Thus, the workshop is the most effective organization for change and, if possible, preferable to the project approach.

Newer project management approaches, such as SCRUM (crowding), which originated in software development, also work with the idea of rapid implementation in small steps. Lean production was an important source of ideas for this now widespread development method. The principles of "flow" and "takt" are the basic principles of SCRUM. The methodology is characterized by the fact that autonomous teams develop the product in manageable and recurring stages, the so-called sprints (e.g., every 14 days). The basis is the common goal, but the path is left to the team.

The topics for improving processes are unlimited – but not the resources for working through them. Therefore, it is important for the project or workshop that the goals, but also the necessary resources, are clarified in advance and considered in the plan. You can then see whether the focus is set correctly, not too many points are made in parallel and the resources are sufficient.

Change is mainly planned and managed in workshops or projects. Nevertheless, even small improvements that are identified in everyday life can be implemented quickly and unbureaucratically by the company's own employees. These quick wins bring about an operational improvement, but above all they are also motivating for the employees. They are mostly improvements that employees and customers feel, even if they are not always measurable in terms of financial results.

In a company with around 70 production employees, the launch of a comprehensive 5S program enabled over 150 improvement initiatives to be identified in 12 weeks, and around 100 of these initiatives were implemented directly without a major project. In total, this company saved around $300,000 within 12 weeks. What a success.

Plan From the Inside Out and Upstream

The pace of change is determined in workshops and a overall project set up and plan. It is crucial that you do not overload the plan and do not change everywhere in parallel, but area by area, process by process. But in which area should these changes start, with which topics and in which order?

Certainly, there are situational conditions that determine an ideal sequence, for example, if an area has a particularly high level of waste or the process is a bottleneck. In principle, however, change from the inside out and upstream makes sense.

From the inside out means that you first start the improvement in the value-added areas, then you tackle the material supply of these areas. First the internal supply logistics, then the external. This is followed by the indirect areas such as purchasing, sales and development.

Upstream means that improvement starts in shipping and then continues, process by process, from final assemblies to single-part production toward value-added processes.

This customer- and value-oriented approach prevents waste from being swept out of areas and brought back in through the back door from other areas.

The sequence of Lean methods also follows this logic. For example, you can lay the foundations with 5S in the value-adding areas and then ensure takt and flow principles here (for instance, using the methods line balancing and zoning). The next step is to improve the lower value-added levels (method SMED) and the material supply (using the methods Milkrun,

Kanban, sequencing). In parallel, attention must always be paid to process stability (A3, Andon, shop floor management).

Plan and Build Resources

You cannot make the transformation to a Lean company without committing financial and human resources and time. Therefore, the commitment and long-term will for change from management is crucial, because management must provide these resources and make sure that the priority for change is not lost in the day-to-day operations. We have already discussed this in detail at one point or another. Investment decisions for resources show the real will of management. That is why the honest commitment to the change is evident at the latest now, when the question of resources arises.

There are several ways to make resources available for projects and bring change forward. The example of the Swiss bathtub manufacturer shows how deliberate free spaces for change can be created. The employees and teams should achieve defined goals in the workshops. For this purpose, they must be released from other activities during this time.

External Lean consultants can bring Lean experience from many companies and provide good start-up support, especially at the beginning of the change. External challenge is often needed and helps in focusing on the right path. For larger companies (over 150 employees), it will also quickly pay off to have their own Lean organization, headed, for example, by a Lean manager who brings methodological expertise and moderates the change process. In any case, over time, employees from each department should be empowered to lead the Lean projects. As Lean activities increase productivity, capacities and employees are freed up. As Lean multipliers, these employees can further advance the change. In this way, capacity for change can be created from the improvements that have been implemented. This is a good opportunity to motivate and develop employees from the operational area.

5.7   Step 5: Implement Change

A Gap Between Knowing and Doing

The change to Lean takes place 90% in the heads, but 100% on the shop-floor. The will to change, the knowledge of how to make the change happen and a good plan are three prerequisites for achieving the results. But you ultimately must make the change happen. That means reengineering processes and changing things visibly in the shopfloor and not only on paper. After all, the effort will not be cost-effective until a physical and visible change has taken place. That is, when the shelf has been moved closer to the espresso machine.

Implementation gains momentum when everyone pitches in. Management must also be involved in the concrete implementation, even if the agenda is full. Participation in workshops must also become a duty for the boss. For specific implementation workshop to be effective, the means for implementation must be there to build up the shelves, tools, scales and fixtures. The most effective ways to implement ideas in the workshop is to organize a company own Lean workshop area, such as a specific maker space to create the needed changes.

The Lean Workshop Maker Space

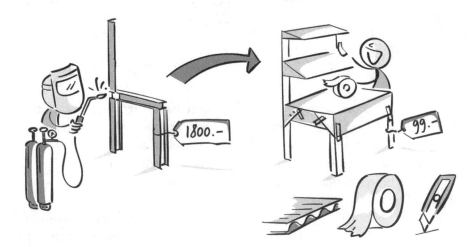

With a Lean workshop maker space you can produce work equipment according to need and so, change can speed up. Here, ideas can already be implemented in the workshop by the team. We have found that small technical tools or even rebuilds are critical for quick success.

PowerPoint graphics can show the way and the concept, but their physical implementation can eat up a lot of time. That is why internal expertise and capacity for concrete technical implementation can be critical. It is important not to overlook testing these tools when making many changes. Only in this way can problems be intercepted in advance. Be it operating equipment, shelves, transport carts or entire workstations. In the Lean workshop maker space, functional prototypes of these solutions can also be produced quickly.

For example, cardboard engineering can be used to quickly and inexpensively map a 1:1 model of the workplace to realistically test the new processes. Like a 3D printer for physical products.

Errors in the concept are thus easily and immediately detectable before major investments are realized.

Control the Implementation

The implementation of the change plan must be steered after it has been adopted. Regular steering meetings accompany and steer the change. In this way, the workshop and project results can be presented to the board by the company's own employees, if possible, with quantified results. This gives employees the trust that implementation is important and is wanted and supported by management.

However, the steering meeting is not a one-way street. Employees should present the results of workshops and projects, but also ask for help when problems arise. In the sense of feedback or an expert opinion, the teams should thus be helped to overcome the hurdles that cannot be overcome within their own team.

In the spirit of Lean, the steering meeting should take place periodically according to a fixed rhythm. Every efficient meeting needs a communicated agenda and should be short – certainly not longer than one to two hours. In addition, the management should participate in the meeting.

In every workshop and project, the results should be measured and documented. In particular, before and after photographs of the change help to show the successes and motivate other areas. Completing initiatives with measurable successes also increases acceptance and the will for further change.

5.8   And What Happens Next?

We are slowly coming to the end of our Lean journey. We hope we have inspired you to see processes with a different eye – >the Lean eye. You are now a real Lean and process expert and designer. And we are sure you already have many ideas for your change. So now let us go and streamline your processes!

We have put the focus in our book on Lean in production. But you can also apply Lean outside of production. The spaghetti diagram can be used to analyze routes to the copier and to documents in the office, and the swim lane char can also be used to bring waste to light in purchasing or sales processes. And equally, the development process can be trimmed to Lean. Lean development is the buzzword. Principles such as cycle and flow are just as important in development as they are in production. Kanban, Shopfloor Management or A3 can also be applied here, with just a few adjustments.

And Lean does not stop outside the factory either. More and more, Lean is being applied in a wide variety of industries. In construction, we speak of Lean construction, in healthcare of Lean hospitals, in software development of Lean software development or even in agriculture of Lean farming. These are just a few of the industries that have recognized the potential of optimizing processes with Lean approaches and are systematically using Lean principles and methods to increase efficiency. In many industries, such as construction, traditional Lean methods have been supplemented and new, specific methods have been developed. But there is more to come. Schools and universities can also learn a lot from Lean and streamline their processes internally. Many processes here are administratively driven and often not aligned with customer benefits. A lot of money and time is still wasted here. Lean is therefore continuously making its way into many areas. This is also true for the government, also Lean is definitely a key to better and customer-oriented processes.

The content also continues to evolve. The Lean philosophy and Lean methods are constantly evolving. A current and large area of research in this context revolves around the question of how digitalization can support Lean. New technological approaches promise great potential for industry in this regard. Everyone is talking about Industry 4.0, and companies are trying to find their digital way. What does this mean for Lean? Will Lean experience additions or even changes? We see many good approaches and products from the field of digitalization that make Lean easier, faster and more efficient to implement. From the intelligent machine to the digital shop floor management to the driverless milk run with intelligent route, the possibilities are almost limitless.

No matter what level of Lean you are at or what industry you are in. Avoiding waste never ends. So, it always starts all over again with the question: ***Where can I still find and reduce waste? ... also applies at home, by the way.***

References

Imai M (1986) Kaizen (Ky'zen): Te key to Japan's competitive success. McGrawHill, New York

Liker JK, Ogden TN (2011) Toyota under fre: Lessons for turning crisis into opportunity. McGraw-Hill, New York

Ōno T, Bodek N (2008) Toyota production system: Beyond large-scale production, [Reprinted]. Productivity Press, New York, NY

Ōno T, Hof W, Stotko EC, Rother M (2013) Das Toyota-Produktionssystem: Das Standardwerk zur Lean Production, 3., erw. und aktualisierte Auf. Produktion. Campus-Verl., Frankfurt am Main

Rother M, Shook J (2009) Learning to see: Value-stream mapping to create value and eliminate muda, Version 1.4. A lean tool kit method and workbook. Lean Enterprise Inst, Cambridge, Mass.

Schönsleben P (2016) Integrales Logistikmanagement: Operations und Supply Chain Management innerhalb des Unternehmens und unternehmensübergreifend, 7., bearbeitete und erweiterte Aufage. Springer Vieweg, Berlin, Heidelberg

Syed M (2015) Black box thinking: Te surprising truth about success (and why some people never learn from their mistakes). John Murray, London

Wikipedia (2020a) Prinzip. https://de.wikipedia.org/w/index.php?title=Prinzip&oldid=196498109. Accessed 28 September 2020

Wikipedia (2020b) Verschwendung. https://de.wikipedia.org/w/index.php?title=Verschwendung&oldid=203991540. Accessed 28 September 2020

Womack JP, Jones DT, Roos D (2007) Te machine that changed the world: Te story of lean production ; Toyota's secret weapon in the global car wars that is revolutionizing world industry, 1. pb. ed. Business. Free Press, New York, NY

© Springer-Verlag GmbH Germany, part of Springer Nature 2022
R. Hänggi et al., *LEAN Production – Easy and Comprehensive*,
https://doi.org/10.1007/978-3-662-64527-7